AF865566

Research for Development

Series Editors

Emilio Bartezzaghi, Milan, Italy

Giampio Bracchi, Milan, Italy

Adalberto Del Bo, Politecnico di Milano, Milan, Italy

Ferran Sagarra Trias, Department of Urbanism and Regional Planning, Universitat Politècnica de Catalunya, Barcelona, Barcelona, Spain

Francesco Stellacci, Supramolecular NanoMaterials and Interfaces Laboratory (SuNMiL), Institute of Materials, Ecole Polytechnique Fédérale de Lausanne (EPFL), Lausanne, Vaud, Switzerland

Enrico Zio, Politecnico di Milano, Milan, Italy
Ecole Centrale Paris, Paris, France

The series Research for Development serves as a vehicle for the presentation and dissemination of complex research and multidisciplinary projects. The published work is dedicated to fostering a high degree of innovation and to the sophisticated demonstration of new techniques or methods.

The aim of the Research for Development series is to promote well-balanced sustainable growth. This might take the form of measurable social and economic outcomes, in addition to environmental benefits, or improved efficiency in the use of resources; it might also involve an original mix of intervention schemes.

Research for Development focuses on the following topics and disciplines:

Urban regeneration and infrastructure, Info-mobility, transport, and logistics, Environment and the land, Cultural heritage and landscape, Energy, Innovation in processes and technologies, Applications of chemistry, materials, and nanotechnologies, Material science and biotechnology solutions, Physics results and related applications and aerospace, Ongoing training and continuing education.

Fondazione Politecnico di Milano collaborates as a special co-partner in this series by suggesting themes and evaluating proposals for new volumes. Research for Development addresses researchers, advanced graduate students, and policy and decision-makers around the world in government, industry, and civil society.

THE SERIES IS INDEXED IN SCOPUS

Alice Buoli · Oana Cristina Ţiganea
Editors

Territorial Fragilities in Cyprus

Planning and Preservation Strategies

Fondazione Politecnico di Milano

Springer

Editors
Alice Buoli
Department of Architecture and Urban Studies (DAStU)
Politecnico di Milano
Milan, Italy

Oana Cristina Ţiganea
Department of Architecture and Urban Studies (DAStU)
Politecnico di Milano
Milan, Italy

ISSN 2198-7300 ISSN 2198-7319 (electronic)
Research for Development
ISBN 978-3-031-36075-6 ISBN 978-3-031-36076-3 (eBook)
https://doi.org/10.1007/978-3-031-36076-3

© The Editor(s) (if applicable) and The Author(s), under exclusive license to Springer Nature Switzerland AG 2023

This work is subject to copyright. All rights are solely and exclusively licensed by the Publisher, whether the whole or part of the material is concerned, specifically the rights of translation, reprinting, reuse of illustrations, recitation, broadcasting, reproduction on microfilms or in any other physical way, and transmission or information storage and retrieval, electronic adaptation, computer software, or by similar or dissimilar methodology now known or hereafter developed.
The use of general descriptive names, registered names, trademarks, service marks, etc. in this publication does not imply, even in the absence of a specific statement, that such names are exempt from the relevant protective laws and regulations and therefore free for general use.
The publisher, the authors, and the editors are safe to assume that the advice and information in this book are believed to be true and accurate at the date of publication. Neither the publisher nor the authors or the editors give a warranty, expressed or implied, with respect to the material contained herein or for any errors or omissions that may have been made. The publisher remains neutral with regard to jurisdictional claims in published maps and institutional affiliations.

This Springer imprint is published by the registered company Springer Nature Switzerland AG
The registered company address is: Gewerbestrasse 11, 6330 Cham, Switzerland

Contents

Introduction: Crossing Borders/Building Bridges. An Interdisciplinary and Research-By-Design Approach to Nicosia's Territorial Fragilities

Alice Buoli and Oana Cristina Ţiganea

Abstract This introductory essay aims to frame the book's primary research setting, contents and contributions with reference to a "polytechnic" interdisciplinary approach to studying fragile contested territories and cities in the Euro-Mediterranean space. The main research background and premises linked to the inter-doctoral research initiative entitled "Territorial Fragilities in Cyprus" are introduced, making evident the relevance of a trans-cultural and research-by-design approach towards a sensitive urban and built environment strongly politicized for its patrimonial acknowledgement while abandoned and defined by advanced material decay. Furthermore, the issues of knowledge production, urban design and planning, and architectural preservation within a divided territory are presented. The "bridging" communication and collaboration potential of the built environment is introduced to redefine local uses and urban functions. The case of Nicosia, and Cyprus more in general, is illustrated as a critical contest to explore integrated planning, conservation practices and tools in areas of historic urbanization crossed by (political) borders.

1 The Role of Language and Space in Contested Territories

Over the past three years, several major dramatic events have occurred, resulting in irreversible consequences for how we live and perceive our cities and homes, as well as the safety of public and domestic spaces. From the global Covid-19 pandemic in 2020 to the Russian invasion of Ukraine in 2022 and, more recently in 2023, the devastating earthquakes sequence in Southern Turkey and Syria. These events have widely prompted a deep reflection on the resilience and adaptability of local communities in the face of major political and environmental crises and their impacts

A. Buoli (✉) · O. C. Ţiganea
Department of Architecture and Urban Studies, Politecnico di Milano, Milan, Italy
e-mail: alice.buoli@polimi.it

O. C. Ţiganea
e-mail: oanacristina.tiganea@polimi.it

© The Author(s), under exclusive license to Springer Nature Switzerland AG 2023
A. Buoli and O. C. Ţiganea (eds.), *Territorial Fragilities in Cyprus*, Research for Development, https://doi.org/10.1007/978-3-031-36076-3_1

across different scales on territories and societies, exposing the urgent need for multi-disciplinary and trans-scalar approaches to address and rethink such issues.

In this context, Cyprus, an island located at the "threshold" between Europe and the Middle East, and at the intersection of major (geo)political and geological fault lines, provides a unique vantage point from which to observe and respond to these complex challenges. It further represents an observatory of more resilient and sustainable approaches to urban planning and architecture while tackling the island's most pressing territorial fragilities. This is particularly evident if we look at the case of Nicosia, one of the divided cities in the world and the capital of both a member state of the European Union since 2004 (the Republic of Cyprus—RoC) and the Turkish Republic of Northern Cyprus (TRNC). For more than five decades now, the Green Line, an "open wound" in the geopolitical body of Europe, cuts through the whole island of Cyprus as well as through Nicosia, including its historical center surrounded by the sixteenth-century Venetian Walls.

As a consequence of the anti-colonial struggles, ethnic and geopolitical turmoil that occurred in Cyprus since the 1960s (and more permanently after the events of 1974), the Green Line separates materially and symbolically the island into two distinguished territories and communities (Turkish-Cypriot in the North and Greek-Cypriot in the South). The roots and reasons for the partition are extremely complex and controversial and reflect very diverse and "entangled"[1] positions, identities, and imaginaries. It is not our intention to explore in depth the historical antecedents and roots of the "Cyprus problem".[2] However, it is worth mentioning here how the presence of the partition became engraved in space, language and people's imaginaries while it has been impacting everyday life practices and spaces of all Cypriot communities, including the foreign citizens that in recent years have settled down on the island.

"Occupation/occupied territories", "border", "displacement", "refugees", "enclaves", "us/them", "property" and "restitution" are only some of the most recurrent words that emerge from conversations with different generations of Cypriots on both sides of the divide. A recent initiative entitled "Cyprus Media Dialogue Project" funded by the OSCE and involving journalists from both communities as well as international colleagues, focused on the role of language and communication in the island aiming to "contribute to media pluralism and free flow of information, and advance accountable, quality journalism".[3] One of the main products of the project is a glossary, entitled "Words that matter: A Glossary of Journalism in Cyprus" (Azgın et al. 2018), including some of the most recurring expressions and phrases commonly used by the two communities. Interestingly, according to the study, words that have a spatial connotation can assume very diverse meanings,

[1] "Entangled Milieus" is the title of one the research projects developed by prof. Socrates Stratis within the activities of LUCY (Laboratory of Urbanism at the Department of Architecture of the University of Cyprus) in cooperation with TU Delft and the University of Sheffield.

[2] For more detailed and clear explanation please see James Ker-Lindsay (2011), Strüver (2018), Papadakis (2018) and Casaglia (2020) among others.

[3] Source: https://www.osce.org/representative-on-freedom-of-media/cyprus-dialogue. Accessed April 2023.

strongly politically charged, depending on the person or community using the same expression. For instance, when referred to the Green Line, the term "border" is considered differently by the Turkish Cypriot or the Greek Cypriot community: "there is a Greek Cypriot view that using the word 'border' as an indicator of geographic boundaries of political entities or legal jurisdictions in the Cyprus context may suggest that there are two sovereign states in the island separated by a border" (Azgın et al. 2018). Similarly, Papadakis (2018) explains the "paradox" and the ambiguity of Cyprus' border in historical, cultural and political terms which are "perceived in different ways by people in each side: Turkish Cypriots more often argued it was a line of protection, a state border and a permanent division, while Greek Cypriots regarded it as a line of aggression, a boundary with an unrecognized political entity, and an impermanent one that should be abolished in order to reunite the island" (2018: 288).

Along with spatial expressions of division or occupation, as the Turkish Republic of Northern Cyprus (TRNC) is not recognized by the Republic of Cyprus (RoC) and by other countries, except Turkey, all terms related to institutional forms or collective infrastructures, can become objects of contestation, depending on who is using them.

However, both in the RoC and in the TRNC, the built environment and the social interactions and practices occurring in space and through collective infrastructures, like those in Nicosia, are expressions of a *civitas*, bonded by rules of coexistence and governed by elected representatives.

As the essays in this volume will show,[4] language and space contribute to the formation of collective and individual identities that are constructed through the exposure to verbal and non-verbal / tangible and intangible expressions of belonging that can either align or collide with diverse word views and narratives. However, as language, space can be molded, bended, and "manipulated", conveying diverse messages and symbolic meanings, including ways of re-conceiving the partition (and its various articulations along and within the Buffer Zone) away from binary oppositions and expressions.

The manipulation of historical built heritage to define specific local, regional, and national identities is something quite common throughout modern and contemporary history, when the continuous redrawn of geopolitical maps and circles of influences (i.e., religious, economic, cultural) hardly takes into consideration the connection between space, identity, and community. Furthermore, the manipulation of cultural heritage in demeaning and even canceling "the other" became quite in the spotlight of all recent armed conflicts and wars. Less attention has been given to the situations in which elements of the cultural heritage (tangible and intangible) brought together communities in dialogue and communication. As "space matters" when dealing with local memories and identities (Nora 1989), the role of heritage as peace-making device can become the most desired tool in "stapling" what was "broken" by the partition. This relates also to the effects of the partition on the fragility of the built environment dealing, both in south and north Nicosia, with the same construction

[4] See for instance the contribution by Montedoro and Lemes de Oliveira.

materials and techniques and, thus, experiencing similar problems when dealing with its vulnerabilities.

Among the two communities is shared a common preoccupation for the state of conservation of the cultural heritage within the Buffer Zone, something already tackled by specialists from both sides in Nicosia, that organized common surveys and proposed interventions.[5]

Furthermore, the state of abandonment and decay of the built environment outside and in close proximity to the Buffer Zone keeps citizens and members of the communities afar while attracting new populations that perceive and interact differently with the existing urban and architectural heritage. On top of that, we should mention the commonly shared values of the cultural heritage, such as the Venetian Walls, which represent a symbol of Nicosia as a whole. The overlapping of these spaces and symbols, elements of the local cultural heritage, has the potential of enhancing communication and cooperation by tackling everyday urban habits, practices and *new* uses of the built environment.

The research presented in the volume explores alternative views from the North–South divide, to suggest—through the "external" viewpoints of a group of affirmed and young scholars—a combination of research-by-design, place-based, and policy-oriented approaches to urban transformation and heritage preservation, in which tangible and intangible elements are equally considered and synergically reframed into short- and long-term scenarios. The theoretical reflections and (meta)design explorations presented in the essays suggest the inter-communal use and reuse of the built environment in the proximity of the Buffer Zone, pushing towards its potential permeability and thus opening new opportunities for "visual recognition and social negotiation" (Bakshi 2017: 128) and ultimately of "cohabitation" between all Nicosians.

The idea of bi- (or inter-)[6] communal cooperation in Nicosia comes in continuity with the work already accomplished by the Nicosia Masterplan (NMP) initiated during the early 1980s, which has been one of the main references for the research presented in this book.[7]

As suggested by the NMP team leaders, the original idea behind the program was to plan for the future of the city in light of the potential reunification of the island, but also in view of the current conditions of separation, to offer improved collective services, safe and accessible housing and quality open spaces on both sides, starting from the walled city and the spaces in the proximity of the Buffer Zone.

---

[5] See for example the interview with Güralp and Petridou published in this volume.

[6] As for instance suggested by the approach proposed by the Home for Cooperation, a community centre opened in 2011 and located within the UN Buffer Zone, in the Ledra Palace area. The H4C is regarded by Nicosia's communities as a landmark in the city, "acting as a bridge-builder between separated communities, memories and visions through its physical presence and its peacebuilding programs benefiting from the transformative power of arts and culture". Source: https://www.home4cooperation.info/who-we-are/. Accessed April 2023.

[7] See the essays by Bricocoli and Chrysochou and Gaeta and Pasqui, and the interview with Güralp and Petridou for more detailed discussion.

The research presented in the volume intends to push further the idea of transforming the built environment around the partition, through the involvement of multiple local actors and stakeholders towards the acceptance, preservation, (re)use, and transformation of these spaces while tackling common issues for both south and north parts of the city, and the island in its complexity.

The outcomes of this debate between the *outsiders* (the authors of this book) and the *locals* (specialists invited to attend and actively participate in the project in its different development phases) are integrated within all proposed scenarios for the walled city.

2 Territorial Fragilities in Cyprus and Nicosia

The book builds on the outcomes of an inter-doctoral research initiative among the Urban Planning, Design and Policy (UPDP) and Preservation of the Architectural Heritage (PAH) programs at the Department of Architecture and Urban Studies of Politecnico di Milano. The doctoral project entitled "Territorial Fragilities in Cyprus. Planning and Preservations Strategies" run from February to October 2021 and involved teaching staff and Ph.D. candidates with expertise in the areas of urban planning and cultural heritage in conflict zones, approached from a multi-disciplinary and multi-cultural standpoint.

As editors and authors of this volume—involved from the very early stages of the initiative—we acknowledge the complexity of addressing Cyprus and Nicosia's vulnerabilities emerged as both a long-term consequence of the partition, but also as the combined effects of political, social, environmental and geological threads. Under-used and decaying architectural heritage, sustainable urban development, water scarcity, climate crisis, cultural diversity and access to public amenities are only among the most pressing issues that the essays contributing to the volume adopted as potential common challenges to be tackled synergically across divided territories.

From this point of view, debates on "territorial fragilities" emerged as a general conceptual and cultural framework for the analytical and design-oriented explorations in Nicosia.[8]

We looked at the local fragilities through the lenses of space-society relationship and interconnection while considering Nicosia as a city exposed to different environmental, political, economic, and socio-cultural risks. These are linked to global–local challenges that regard the island in its entirety. This way of reading and interpreting a territory through the effects of major trans-regional processes on the built environment and communities is common to a "polytechnic" approach that was foundational

[8] Since 2018, the Department of Architecture and Urban Studies at Politecnico di Milano has built extensive knowledge and a rich scientific production on the topic of "territorial fragilities" under the "Department of Excellence" program of the Italian Ministry University and Research (MUR) - https://www.eccellenza.dastu.polimi.it/en/homepage/.

to the research, deeply rooted in an understanding of territorial fragilities as potential resources for local sustainable development.

Our "gatekeepers" to Nicosia were the colleagues and experts invited to present their research remotely—also due to the measures against the Covid-19 pandemic that was still in place in early 2021—in the context of an intensive guest lecture seminar organized in February 2021. International contributors included Anita Bakshi, Anna Casaglia, Nasso Chrysochou, Anna Grichting, Ali Güralp, Agni Petridou, and Guido Licciardi, professors and experts in the field of architecture, geography, urban planning and heritage preservation. Lectures touched upon different angles on both the island and the city of Nicosia, such as: the complex geographies of the partition in Cyprus, mapping memories and identities in old town Nicosia, landscape design proposals for the Green Line, the Nicosia Masterplan, the preservation of the architectural heritage of the walled city, urban heritage in fragile and conflict contexts.

This early cycle of lectures and conversations with experts and colleagues allowed both the teaching team and the Ph.D. candidates to develop a framework for their own approach and views on the study context, broadening the fields of analysis and setting the main topics to be later assessed and verified "on the ground" during the fieldwork activity. Due to the pandemic situation, the entire program depended on the mobility restrictions that overlapped with the existing divide in Cyprus. Therefore, after the preliminary research was already accomplished through online lectures, meetings and use of available research tools and databases, the study mission and fieldwork took place in September 2021 and run for a week in south and north Nicosia. Throughout our stay, the involvement of local stakeholders was key for exploring both sides of the city, by including visits and thematic walks curated by our guests. City walks in Nicosia played an important role in both understanding the complex and fragmented geographies of the city, and also as a tool for collaboration: walking together guided by a certain topic or "red thread" represented a powerful experience creating bonds around common interests and views. This approach was pioneered by two of the most important inter-communal initiatives and organizations in the city, the Association for Historical Dialogue and Research[9] and the Home for Cooperation.

Our walks included trails on both sides of the Green Line, touching main heritage buildings and sites in the Chrysaliniotissa neighborhood with Nasso Chrysochou, spaces of nation-building identity in the south with Yannis Papadakis and Michalis Moutselos, North Nicosia public spaces and housing with Bahar Akpinar, Kaimakli neighborhood along the Buffer Zone with Yiorgos Hadjichristou, Urban Gorillas, Marina Christodoulidou and Evagoras Vanezis. Unable to gain access to the Buffer Zone, due to the pandemic restrictions, hindered mobility initially represented a

[9] "AHDR is a unique multi-communal, non-for- profit, non-governmental organization established in Nicosia in 2003. Since its foundation, the AHDR has enlisted members from various ethnic, linguistic, and professional backgrounds working at various educational levels in Cyprus, making the first steps towards a greater effort to maintain a continuous, open dialogue about enhanced pedagogic practices that could encourage the values of the discipline of history". Source: https://www.ahdr.info/.

major limitation of the research program. However, this limitation allowed further investigations of the urban spaces and architectures close to the Buffer Zone, or in direct connection with it, opening new perspectives and scenarios that helped the doctoral researchers to refine and define their ideas, testing them with continuous engagement with the context and the people encountered. The projects eventually focused on all those "residual" spaces (e.g., vacant plots, neglected heritage buildings, living urban ruins, dead ends) in proximity to the Buffer Zone. As a result, the intervention strategies that emerged from the work on the ground suggested reconsidering those places—once positioned in the town center and then turned into peripheral and marginal realities due to the partition—as potential "inter-communal communication bridges" across the city and its communities.

The intermediate results of the research were presented and discussed with colleagues and guests at the Home for Cooperation during the final seminar of the on-site workshop (September 2021). Comments and suggestions by the audience were later implanted in the final report produced by the Ph.D. candidates, that served as a starting point for their contributions in this volume.

Back in Italy, our team was able to engage further with the research results, thanks to an invitation to present the final report at the Cyprus Pavilion at the Venice Biennale 2021, hosted by the curatorial group and in particular Marina Christodoulidou and Evagoras Vanezis. This was a key moment of discussion of the work, including the participation of an "external" audience (the visitors of the pavilion) and other colleagues on-site and online.

The outcomes of the thoughts, talks and walks that occurred in Nicosia, Milano and Venice accompanied the production of this book, which collects theoretical reflections, hands-on experiences and radical design visions for the city in reason of and "beyond the partition".[10]

3 How This Book is Structured

While most recent research projects, design workshops, and/or photographic projects concentrate on the Buffer Zone in its various manifestations and potential regeneration strategies (utopian or not), the book is devoted to analyzing, understanding, and reframing Nicosia's socio-spatial fragilities with the lens of urban planning, design and policy and architectural preservation.

To this purpose the book is structured, besides this introductory essay, in ten contributions by colleagues from Nicosia, professors, and experienced and young researchers from Politecnico di Milano, acknowledging the complexity and variety of understanding, viewpoints and research trajectories developed on the city by the authors.

The first three contributions (Chaps. 2, 3, and 4) aim to provide an interdisciplinary background for the volume, through the voices of scholars and professionals directly

[10] Here the reference is to the book by Anna Casaglia "Nicosia beyond partition" (2020).

engaged with the Cypriot context and its built environment, and through the editors' opening contributions.

Chapter 2 "Moving on the Borderlines. Legacies and Long-term Perspectives of Academic Encounters" by Massimo Bricocoli and Nasso Chrysochou provides a recount of a long-lasting friendship between the two authors, started during an urban regeneration and development workshop in Dortmund when they were architecture students in the early 1990s.

Based on the authors' shared ideas, mutual understanding, and professional and personal knowledge developed over time, the chapter explores the value of on-site didactic activities in borderlines. The two authors engage in a free and open dialogue about the past, present, and future perspectives of the many borders (in space and time) that characterize Cyprus and Nicosia, providing both an academic "overture" and a personal take on the main issues and key topics that will be discussed in the essays presented in this volume.

Chapter 3 "The Experience of the Nicosia Masterplan. Interview with Agni Petridou and Ali Güralp", edited by Alice Buoli and Oana Cristina Ţiganea, collects the results of a conversation recorded in July 2022 with the team leaders of the Nicosia Masterplan. As for the first contribution, the interview recounts crucial turning points both in their professional and personal relationship, another long-lasting friendship that crossed the last forty decades of Nicosia's urban planning and regeneration projects. The interview testifies the role and challenges of being a "public servant" in a contest of a divided city, reflecting the pioneering trajectories, successes but also disillusions of the former NMP coordinators, in view of the current difficulties that bi-communal cooperation is experiencing in Nicosia, marked by political and financial unbalances.

Chapter 4 "Territorial Fragilities in Nicosia. Tangible and Intangible Heritage Constellations in Nicosia" by Alice Buoli and Oana Cristina Ţiganea illustrates the key thematic and methodological elements of the research-by-design didactic explorations proposed in this volume, drawing on and feeding back the theoretical and empirical contributions of the book. Through a series of collective "walks" on both sides of the Green Line—adopted as an explorative method and a narrative device—a series of constellations of open and built public, semi-public and secluded spaces emerge.

Even though discontinuous and episodic, these spatial narratives develop a composite and multi-layered image of Nicosia, of its tangible and intangible urban conditions along and inside the Buffer Zone.

The chapter also introduces the main academic debates on "divided cities", challenging such a definition in the specific context of Nicosia.

The second sequence of contributions (Chaps. 5, 6 and 7) reflects the rich interdisciplinary theoretical dialogues between the research lines and expertise represented in the volume. The contributions discuss the role of urban heritage, planning and design in divided cities with different perspectives and takes on the "Cyprus problem" and on Nicosia's multiple fragilities.

Chapter 5 "Shared "Values" in Divided Contexts. Some Reflections on the Role of Urban Heritage in Cyprus" by Mariacristina Giambruno and Annunziata Maria

Oteri proposes a focus on the role of urban heritage in divided contests. The purpose of the essay is to analyze two different perspectives through a preliminary critical overview of the law and initiatives defined at an international level to protect the cultural heritage in conflicted areas since WWII. The urban and architectural heritage as a peacemaker or, on the contrary, how urban, and architectural heritage can become the symbol of conflicts and divisions. The essay focuses on some specific case studies, analyzing how buildings and sites once considered symbolic and relevant in terms of history, memory, and identities can become "difficult heritage"; the representation of divisions, defeats, and subjugations. In the last part of the essay, a specific reflection on the case of Nicosia in Cyprus is offered, considering the current perception of buildings and sites, once shared values of multi-cultural communities and now divided by a physical, political, but also a cultural border.

Chapter 6 "Engaging Impossibilism in Planning Practice. Reflections from the Case of Nicosia" by Luca Gaeta and Gabriele Pasqui applies the concept of impossibilism (initially developed by Marxist political theorists) to make sense of the bi-communal planning practices undertaken in a city structurally divided along lines of ethnicity and military occupation. The Nicosia Masterplan is a four-decade-long bi-communal initiative to comprehensively plan the city's future, whether political reconciliation will be achieved or never. By pushing the notion of impossibilism beyond the limits of reform under capitalism—as Marxists have it—the authors aim to frame a planning process entailing a radical degree of political uncertainty as part of its mandate. In doing so, the authors are mindful but also critical of Hirschman's possibilism as insufficiently accounting for the Cypriot planners' experience of active disillusionment. Impossibilism describes a condition of political stalemate in which no realistic solution exists for the main problems of urban coexistence. However, planning actions are necessary to make everyday life tolerable. The conclusions discuss tactics and tools for enabling spatial planning when cities are suspended in a state of impossibilism with long-term socio-spatial imbalances.

Chapter 7 "From No Man's Land to Everyone's Space. The Potential of Radical Design to Transform the Green Line in Cyprus" by Laura Montedoro and Fabiano Lemes de Oliveira interrogates the real potential of spatial transformation projects for the Green Line, in light of its political and spatial contentiousness and complexity. How could an intervention in the physical space help foster dialogue between the two parties? And, more broadly, how could the Green Line be re-signified from nobody's land to everyone's space? The operability and effectiveness of design (landscape, urban and architectural) in a situation of long-standing socio-political conflict are deeply challenged; the spaces for interventions capable of making sense are limited.

Starting from the analogy between language and space, this chapter proposes a re-interpretation of the Green Line as a Mediterranean free space, and provocative design actions are hypothesized to reinforce its "pidgin" character as a condition for potential new beginnings.

The third sequence of essays (Chaps. 8, 9, 10 and 11) presents the research focused on Nicosia's old town, through four different urban designs, urban planning and policy explorations, suggesting other ways of seeing and intervening in the city's manifold urban spaces and its cultural, environmental and architectural patrimonial legacies[11].

Chapter 8 "Stapling strategies around the Green Line in Nicosia's Old Town" by Federico Barbieri, Wenshan Chen, Wei Lyu, Francesco Pasta, and Dafni Riga outlines a theoretical and methodological framework for the proposed intervention strategies for Nicosia's walled city center.

The authors define a "stapling strategy" as a process which builds upon and develops existing socio-spatial interlinkages between the two sides of Nicosia's historic city center, metaphorically "stapling" or "stitching" together spaces that used to be contiguous and continuous but are now severed by the Green Line.

The first part of the chapter provides a conceptual background, relating a range of subjects that are relevant to the research and intervention approach in the divided city of Nicosia such as e.g., the concept of a "thick" border, the ambivalent semiotics of urban space, the gap between conceived and experienced space, the materiality of everyday life, and the idea of cross-border engagement. Grounded in the theoretical framework and based on fieldwork conducted in Nicosia, the second section identifies three main themes, which can be considered as both challenges and opportunities from a strategic design point of view. These are: (1) bridging inter-communal planning with civil society; (2) reactivating under-used or abandoned spaces; and (3) building upon a shared socio-ecological system. Such macro-subjects constitute criticalities that present the potential for action and intervention.

The third section outlines the principles of community-based action planning, the key guidelines for intervention, and the community actors to be involved. Based on these elements, the authors discuss how the proposed planning strategies do not constitute a prescriptive masterplan, nor an inflexible vision. Rather, they compose an open framework for strategic and concerted action. They are complementary but not necessarily interdependent.

Chapter 9 "Tackling Residuality through Nicosia's Market Heritage" by Roberta Pellicano, Verdiana Peron, Aubrey Toldi, Constanze Wolfgring, and Shifu Zhang addresses marketplaces and residual spaces in old town Nicosia as catalysts of reactivation strategies. Markets were central in the old city of Nicosia for decades, playing a crucial role in bolstering relations between different communities. Locals developed memories and attachments to these everyday spaces, demonstrating how they are part of a common tangible and intangible legacy, which still serves as a strong point of reference of what Nicosia used to be before the partition. The establishment of the buffer zone in 1964 disrupted the marketspace, transforming once bustling market streets, that were a common ground to the ethnically divided neighborhoods, into disconnected residual spaces. When considering how to mitigate the existing barriers between Nicosians on both sides, these spaces could play a crucial role. Based on the hypothesis that Nicosia's history as a market city embodies a common, predominantly positively connotated heritage, this study explores its potential in tackling

[11] All maps of Nicosia depicting the Buffer Zone use as main reference the "Nicosia Master Plan Archives". Source: http://udsnicosiabufferzone.blogspot.com/p/blog-page_2.html. Accessed April 2023.

residuality within the area of the historical marketspace and in promoting opportunities for dialogue and encounter. The research was developed through bibliographic research, interviews and on-site surveys. The authors propose short and long-term strategies concerning residual spaces and buildings to reactivate and re-center the former marketspace, arguing that these once vivid spaces in the very heart of the city can again assume bridging functions between communities.

Chapter 10 "Civic Water. Bridging Culture, Nature and People" by Samidha Pusalkar, Norma Camilla Baratta, Massimo Izzo, Danila Saulino, and Filippo LaFleur explores the potential reuse of traditional urban water infrastructures in Nicosia, with reference to the walled city. The political and physical partition of the city has not only pushed for the material decay of the built environment inside and bordering the Green Line, but also the abandonment of various common infrastructures, such as the water system. This is even more critical in Cyprus, an island suffering from severe water scarcity. Hence, Nicosia's water legacy has been studied, mapped, and analyzed from its patrimonial perspective, with attention to its tangible and intangible manifestations. This research enables the potential to connect the two communities by virtue of the social and renewable use of water. To promote this vision, the research puts "water heritage" as a theme under which community-based initiatives and the reintroduction of nature as a service in the urban fabric are proposed with different social and cultural target actors, based on the methodology of "stapling strategies".

Chapter 11 "The Venetian Walls of Nicosia. A Palimpsest for a Common Future" by Giulia Bressan, Anna Evangelisti, Paola Martire, and Livia Shamir aims to address the role of Nicosia's Venetian Walls as a platform for symbolic and physical communication bridges between the two divided sides of the city. This goal is rooted in the potential of attributing new meanings to Nicosia's cultural heritage, particularly the tangible heritage of the Venetian Walls. Despite their historical significance in defining the city's history and symbolically representing it, they form a complex defensive system composed of walls, bastions, and moat structures that have always been perceived as shared heritage by the various communities that have inhabited the city since the sixteenthcentury. Even today, despite the persistent division, the Venetian Walls represent a rare example of shared cultural heritage among conflicting communities.

To explore how the Venetian Walls' heritage component can be reconceived from being a valuable static palimpsest into an essential tool and occasion for the construction of new spaces and shared values, this research has developed a multidisciplinary research path and a series of transformative scenarios. This vision aims to reimagine the walls system and its surroundings as new urban green and public spaces, serving as a new urban catalyst that defines the city as a whole. The original contribution of this study lies in the possibility of considering the Venetian Walls system as the object and space where the Green Line can be reconsidered and ultimately crossed, allowing for tangible connections that blend the different urban communities and cultures.

Acknowledgements The editors would like to express their gratitude towards all the colleagues and institutions that have participated directly or indirectly in this book.

The Department of Architecture and Urban Studies of Politecnico di Milano supported and funded the organization of the workshop in Milan and Cyprus, coordinated by Luca Gaeta and Mariacristina Giambruno as part of the activities of the Urban Planning, Design and Policy (UPDP) and Preservation of the Architectural Heritage (PAH) doctoral programs.

We are grateful to all colleagues and friends that contributed to this volume and also those who generously took part in the research activities online and on-site: Anita Bakshi, Anna Casaglia, Anna Grichting, Guido Licciardi, Yannis Papadakis, Michalis Moutselos, Bahar Akpinar, Yiorgos Hadjichristou (Urban Gorillas), Marina Christodoulidou, and Evagoras Vanezis.

We would like to thank the Home for Cooperation who hosted us in its spaces, the municipality of Nicosia for providing support in the production of the maps, the curator of the Cyprus Pavilion at the Venice Biennale 2021, again Marina Christodoulidou, Evagoras Vanezis with Era Savvides and Nasios Varnavas (Urban Radicals).

Thanks to the Italian Ambassador in Cyprus (Andrea Cavallari, 2017–2022) and his staff and family for kindly receiving our delegation at the Italian Embassy in Nicosia.

Last, but not least, thanks to Neofytos Christou who kindly and patiently supported us in the editing of the cartographic materials.

References

Azgın B, Christofides C, Aygin E, Siakalli M (2018) Words that Matter. A Glossary of Journalism in Cyprus. OSCE Representative on Freedom of the Media Office. Available at https://www.osce.org/files/f/documents/3/2/387269.pdf. Accessed April 2023

Bakshi A (2017) Topographies of memories. A new poetics of commemoration. Palgrave Macmillan, Cham

Casaglia A (2020) Nicosia beyond partition: Complex Geographies of the Divided City. Unicopli, Milan

Ker-Lindsay J (2011) The Cyprus Problem. What Everyone Needs to Know. Oxford University Press, Oxford USA

Nora P (1989) Between Memory and History: Les Lieux de Memoire, Representations, Special Issue: Memory and Counter-Memory, no. 26, pp. 7–24.

Papadakis Y (2018) Borders, paradox and power. Ethnic and Racial Studies 41(2), p 285-302. https://doi.org/10.1080/01419870.2017.1344720

Strüver A (2018) Europeanization in Cypriot Borderscapes: Experiencing the Green Line in Everyday Life, Geopolitics 25(3):609-632. https://doi.org/10.1080/14650045.2018.1550390

Moving on the Borderlines. Legacies and Long-Term Perspectives of Academic Encounters

Massimo Bricocoli and Nasso Chrysochou

Abstract Based on the authors' shared ideas, mutual understanding, and professional and personal knowledge developed over time, this chapter explores the value of on-site didactic activities in borderlines. The two authors engage in a free and open dialogue about the past, present, and future perspectives of the many borders (in space and time) that characterize Cyprus and Nicosia, providing both an academic "overture" and a personal take on the main issues and key topics that will be discussed in the essays presented in this volume.

1 Back into the Future. Experiencing Borderlines Through Academic Workshops

1992, Dortmund, Germany. As part of a European program, the Technical University hosted a European Summer Campus dedicated to urban regeneration and redevelopment. A group of 10 selected Master of Science Students in Architecture and Urban Planning from 10 different European countries, were invited to participate in four weeks of intensive sessions of academic seminars and lectures combined with intensive field work. It was in this context that the two authors of this text, as students from Milan and Nicosia, got to know each other, and began to develop a solid and long-lasting friendship and academic collaboration. The "field" was a vacant plot of land in the Nordstadt, an inner city neighborhood north of Dortmund's main railway station. A working class area, with poor quality housing, a very dense urban fabric, shabby and neglected public spaces and lack of green areas. In the recent years, physical decay and social segregation have increased, defining it as a "neighborhood in crisis". At that time, the Nordstadt was part of the "Quartiers en crise"

M. Bricocoli (✉)
Department of Architecture and Urban Studies, Politecnico Di Milano, Milan, Italy
e-mail: massimo.bricocoli@polimi.it

N. Chrysochou
Department of Architecture, Frederick University, Nicosia, Cyprus
e-mail: nassoch@cytanet.com.cy

© The Author(s), under exclusive license to Springer Nature Switzerland AG 2023
A. Buoli and O. C. Ţiganea (eds.), *Territorial Fragilities in Cyprus*, Research for Development, https://doi.org/10.1007/978-3-031-36076-3_2

network, an innovative European Community public policy initiative that promoted partnerships among different European cities facing the problem of urban decline. It was a pioneering initiative that promoted urban regeneration as an integrated process aiming at the physical, social, and economic regeneration of decaying neighborhoods. While at the university we presented and discussed research and projects related to urban regeneration in a variety of European cities, the fieldwork we were involved in literally consisted of digging a pond. With the support of a local community group, we created a small artificial pond in an area that was overrun with cement and mineral surfaces and lacked green spaces. The pond project was supported by young academics interested in advocacy planning and frontline ecological issues.

The group of students discussed the matter very intensely, and some were critical and reacted against the idea of digging a pond and creating an artificial water surface in such an area. However, the arguments put forward by those in favor focused mainly on the values of biodiversity, bringing environmental qualities to a poor neighborhood and investing in a nature-like element that could symbolically open a further discussion. The students did their work, albeit with little motivation, and completed the work in four weeks. The pond was inaugurated with a press conference and some reports in the local media. The students also engaged in the academic side of the program, writing, and presenting papers and contributing to discussions some of which led to further collaborations and research projects.

After three decades, this experience still looms large in our memories and in the discussions and collaborative academic work between the two of us. We recognize and experience the importance of this workshop played for our individual perspectives and trajectories, both in terms of content and in terms of personal and cultural development in our approach to urban issues. Moreover, a solid and long-lasting friendship is a side effect that any good academic program might wish to include among its merits and possible—but not programmed—outcomes.

To have a group of qualified and selected students digging a pond in a poor neighborhood was at that time a very "frontier" idea, especially from a Southern European perspective, like our Cypriot and Italian ones, where the reference to environmental and ecological issues in urban planning and architecture or to re-naturalization was rather beyond the approaches being experimented in Germany at that time, for example in the IBA Ruhr program.

Culturally, it was an intentional exposure to broaden our understanding of urban change by introducing factors—i.e. nature—that were not usually part of our cultural and professional frameworks. After several years we can recognize how this weird field work was anticipating and dealing with issues and topics that are now largely mainstream: nature-based solutions, biodiversity, and urban regeneration policies that address social and physical segregation.

2 Borderlines in Urban Space and Planning

Somehow, this first experience has subtly inspired and influenced our future choices and academic programs. It was a chance to see how academic work can be effectively developed by combining desk research and fieldwork, enriching research with empirical experience, local knowledge and qualitative research. Obviously, this experience was an implicit reference and basis for the development of a similar program which we coordinated and managed. We developed a concept for a series of three intensive workshops, developed with five partner universities, under the title "Borderlines in urban spaces and planning" which was funded by the European Union under an Erasmus-related project. The program envisaged that thirty selected Master's students in Architecture, Urban Planning and Sociology (from Cyprus, Germany, Italy, and Romania) would participate in an intensive workshop to be held in a different location each year. The sites were chosen according to the concept of exploring borderlines: the remote and peripheral brownfield of the former Aspern airport in Vienna, the former harbor area in the inner city of Hamburg, and the former Paolo Pini psychiatric hospital in Milan.

The three selected cases were somehow sites that had been segregated for decades into monofunctional uses, cut-off from urban development, leading to an exclusion from urban life and defining precise boundaries and perimeters that literally could not be crossed. Despite being involved as a partner university, Cyprus was not included as a case study, while the presentations we had at each workshop discussed the case of the divided inner city of Nicosia and the overall situation of the island, as a reference case.

Organizing the workshops and guiding the students in the understanding of each situation within a general conceptual framework was a very intense experience. The three areas were somehow cases of land use change, of redevelopment, without a lot of political, symbolic, or conflict-related implications. The students and the teaching staff were constantly confronted with the questions and somehow the 'wonders' of what possible destiny and perspective these sites could have, how urban planning and architectural design could support, and shape change within processes that would take decades, a lot of imagination and governance and management skills to implement. As we have already experienced in Dortmund, one of the main challenges is to create the conditions for students to be open and creative in designing possible development scenarios that at the moment seem impossible. While architecture students tend to focus on creativity and fantasy, planning students tend to care for reason and feasibility. In both cases, however, developing a realistic vision of a coherent change in an urban area is a great and challenging exercise.

After several years, we can acknowledge that the Seestadt Aspern is now a fast-growing new neighborhood in the expansion of Vienna, the HafenCity is a very outstanding urban addition to the Hamburg city center, the former Paolo Pini is a wonder of vibrant and multipurpose cultural, entrepreneurial and health projects (Bricocoli and Breckner 2012; Bricocoli et al. 2015). From our perspective and shared experience, the 2021 workshop held in Nicosia was an ideal follow-up and mature step of a series of initiatives in which we have observed how architecture and planning education can benefit from addressing issues of urban change in the face of borders. The old town of Nicosia and the North–South divide represent and add themes to the set of complex elements that we have been dealing with in our previous and shared experiences.

Our contribution to this volume essentially conveys some of the incredible flow of thoughts, shared ideas, mutual understanding, and knowledge that has developed year after year in a professional and personal dimension. A loose and open dialogue between the two of us opened to an intensive exchange of our view on the past, present and perspectives of the borders that characterize Cyprus. We may try to share with the readers some of the views and insights that we have jointly gained together through discussion and exchange.

3 Architecture, Urban Planning in the Face of Time and Long-Term Perspectives

From our personal experience, an important factor for a discussion and any research work on the issue of areas where development processes are intertwined with issues of borders is strongly related to a question of time. Practices define borders (Gaeta 2018; Damgaci and Dağli 2018) and borders define identities and approaching, crossing, or even dismantling them in order to define (invent) new ones requires an extraordinary work of collective elaboration, of multiple perspectives, negotiations and agreements. This is the case for a secluded inner area in search of qualification and integration into the urban fabric of Dortmund, for a no-go area such as a former harbor area with its rigid infrastructural and physical boundaries that need to be overcome if it is to have any prospect of being integrated into the city center, for a distant airport compound located not far from the former Iron Curtain that is being targeted as a new residential neighborhood. It is even more the case of a former psychiatric hospital, literally a black hole in the life of a city, completely isolated from the outside, which is going through a process of opening up that requires an incredible effort both to provide good living conditions for the former inmates and to overcome the stigmatization of the place, while the project is to develop it as a champion of innovative health and social policies based on cultural activities. The case of areas where borders are linked to conflicts is further complicated by the whole range of traumas involved.

Time is relevant in many ways. It is relevant to those who are individually and personally involved and who need time and processing to develop a fuller perspective on issues involved, to overcome traumas and wounds, to experience a multitude of different points of view and to gain a more articulated perspective. Time is important for those who have an interest—whether a personal or research one—and are looking for clues, guides, experts, and witnesses who very often need a relationship of trust to open the doors to the field and to offer their knowledge.

We have experienced this in Vienna, a city that itself only began to recover and develop into a vibrant growing city, only decades after the Second World War and which for a long time was on the borders of Western Europe and so close to the Iron Curtain that any investment was held back for years.

Our common reflection is that the attitude of architects and planners is very often to explore urban and regional contexts with a very strong project-oriented attitude. This often leads to the expectation that major urban projects and the changes they bring can be implemented within the short time frame of a political term or of the individual expectations of a professional architect/planner. When we think about the problems of a divided city and country, the perspectives need to be widened and broadened. The reference of plans and projects must extend into the past as well as in the future and this requires time and patience. It is the time of history and of political decisions but also the time of cultural development and collective awareness which is necessary to reveal not only interpretations but also openness to a sensitive approach and access to relevant information. Developing long-term relationships with places and people in order to gain sensitivity and the ability to deeply understand local conditions and issues, is a fundamental factor in our understanding. While discussing about the Nicosia workshop and the contribution to this book, after more than twenty years, we have had the opportunity to discuss with some older friends how much trust and confidence are needed as a terrain for opening up to memories that can really help to develop a comprehensive understanding but that also recall individual traumas and troubled times.

4 Only Now…(Time and Trust)

In the early 1990s—when he first visited Nasso in Nicosia—Cyprus seemed to Massimo as the most cosmopolitan place he had ever experienced. Everybody had studied abroad: the United States, England, Greece, France, Italy. Everyone in our age group had graduated from a university abroad and very often had received a scholarship for further studies or a first job. This was generalized, as there was no significant presence of universities in Cyprus. Several of the expats who were temporarily appointed in Cyprus, remained on the island for long.

Now, as we sat in Nicosia in the summer of 2022, sharing and discussing the content of the Nicosia workshop and the overall work of the PhD candidates from the Politecnico di Milano, the dear 90-year-old friend Andreas began to recall his own story and we began to record his words as they gave a vivid perspective to our writing.

The sense of cosmopolitanism that pervaded the island in Massimo's memory has had even more extraordinary heights in Andreas's account. He recalls his first big trip out of Cyprus when—at the age of 18—he won a scholarship to attend a major art school in London at the time of the British jurisdiction. In those days—the early 1950s—it that was an amazing journey: traveling by boat to Athens, then to Naples, then to Genoa, then by train to Paris and Calais and then by boat to

Dover and finally by train to London and to the art school. Traveling and stretching visions in a multilocal dimension in a perspective that was not that of migration, but of pioneering multilocality. After completing his studies in England, Andreas would return to Cyprus to work in the public administration and as a renowned artist, maintaining a close exchange with England as well as with many other countries around the world. For many Cypriots, studying abroad has been a necessity and an amazing experience, broadening their horizons while cultivating and nurturing the attachment to their roots. The establishment and opening of universities on—both sides of—the island has changed this perspective in a profound way and there is much to learn and discuss about its implications. The university programs are aimed at local and international students, taking advantage of the island's strategic location in the Eastern Mediterranean and close to the Middle East, and cooperating within the EU academic framework.

We take advantage of the vivid memories and frank conversation to talk to Andreas about relations on the island. As it is generally known, 1974 is a key date that marks the beginning of a division of the country and the city of Nicosia after the Turkish invasion following the attempt of a coup and the political turmoil in Cyprus. This key reference to 1974 turns out to be a formal reference in the discussion of the partition and division of the city center, while the Green Line is assumed to be a buffer that was somehow won and imposed, to allow the two communities to coexist closely in the city center Andreas explains to us how, in fact, a process of segregation was already underway in the historic city center in the 1960s. Due to instability and unrest, Turkish Cypriots had begun to concentrate in the northern part of the city center and Greek Cypriots in the southern part. This was also because it was in the northern part of Nicosia that most of the buildings from the Latin period happen to have survived. Apart from the old Augustinian Monastery of St Mary, now the Omeryie Mosque, there are no significant Gothic buildings in the Greek Cypriot part of Nicosia. The abandonment of all Latin religious buildings and the Latin aristocratic houses in 1571, when the Latins were forced to flee the island after the Ottoman takeover, and their replacement by the Ottoman population resettled on the island, meant that the city's population had already shifted to the current position of the two main ethno-religious groups, Turkish Cypriots to the north and Greek Cypriots in the south. Latin cathedrals and churches which were built north of the river were converted into mosques from the end of the sixteenth century. Of course, bathhouses were built around them for Muslim absolution before prayers and *caravanserais* and *hans* as well as markets to accommodate the pilgrims from the countryside. Before and even after it was diverted out of the city, the river already marked a spatial division in the city. Even after the riverbed had dried up it was not until the beginning of the twentieth century that it started to be filled in with buildings. The river became a street which has survived to the present day. The isolation of the Turkish population in enclaves in the 1960s and the precarious situation arising after the 1974 partition meant that the development of the northern part of the city stagnated. This is a relevant information which has not yet been considered, and which did not correspond to any official or administrative partition, but which had significant effects both on the spatial organization of the two communities and on the urban change as a whole. In a way, this process marked and confirmed the division of the city, whose morphology is determined by the riverbed. But more than that, this movement of the two communities allows us to interpret the different built environment that we can still see today in terms of the different characteristics of the two sides: in the south side an incremental replacement of the historic building stock and low-profile retail and offices, in the north side a slower pace of change, with persistent historic housing and structures.

5 Update. Walking Down Through Old Nicosia

The first steps towards the revitalization of the walled city and the first financial initiatives for the restoration of historic buildings began in 1979, with the aid of the United Nations Development Program (UNDP) and the United Nations Center for Human Settlements (UNCHS). Policymakers from both parts came together to organize a plan aimed at the development of a potentially united city and its revitalization. Under the leadership of the Greek Cypriot Mayor of Nicosia Municipality Lellos Demetriades, and the Turkish Cypriot Mayor of Nicosia Turkish Municipality, Mustafa Akinci, the municipal authorities first worked together in 1978 to design and implement the Nicosia sewage system for the entire walled city. Soon a sophisticated master plan for Nicosia as a single entity was developed. The second phase of the Nicosia Master Plan (1984–1985), focused on the objective of housing and infrastructure and the general improvement of an atmosphere of mutual trust, confidence and respect between the Greek Cypriot and Turkish Cypriot communities (Abu-Orf 2005; Ewers 2018).

The first crossing to open after the division of 1974 took place in April 2003 after almost 30 years of division. This was the Ledra Palace crossing just outside the walls of old Nicosia which did not allow the understanding of the city as one. The subsequent crossing of Ledra Street on the other hand allows for a smoother transition, directly linking the two commercial areas of the walled city.

The move of the Department of Architecture of the University of Cyprus into the old town in 2017 gave the area its first revitalizing boost. Together with the Frederick located to the east of the old town, the two local departments of architecture have encouraged more independent research into the old town. However, most students who study in Cyprus tend to live at home and although the university students have promoted youthful functions, there is still no affordable student housing in the old city.

The financial crisis of 2010, the need for more affordable rents and facilities for young people, combined with the increasing settlement of immigrants in the old city in search of cheap housing in dilapidated and squalid conditions, has driven out investment in the more upmarket facilities and thus the potential for a greater mix of uses. The buildings along the Green Line are another unresolved issue. This forgotten heritage has owners who may be long gone and buildings that are rapidly deteriorating, damaged by military occupation as well as looting and general neglect.

A move to restore the facades of buildings along part of the Green Line was a good cosmetic move but unfortunately a very temporary solution.

Today, the trend is to convert larger buildings and subdivide them into one-bedroom apartments or studios and to densify and fill in empty courtyards. According to Andreas, this trend has the potential to be disastrous for the original typologies and morphologies of the built spaces. This continuous densification which began already in the colonial period but has continued and increased exponentially, has a direct impact on the environment. In the sixteenth century, when Savorgano was building the new walls, he wrote in his diaries that within the new walls, there was enough space to grow wheat which, could feed the inhabitants of Nicosia for a year in case of a siege. Kitchner's map of 1882 also shows fully planted inner courtyards and large areas of the city still under cultivation. The continual sealing of the ground to accommodate extensions, parking lots and seating areas in commercial outdoor spaces is changing the bioclimate of the walled city into an arid space with very little vegetation (and the potential for flooding).

Some qualified restoration interventions have taken place (some contemporary art, crafts, student housing) but a very large part of the building stock consists of 2–4 storey buildings dating from the 1950–1960s, housing, retail, warehouses, offices and of very low quality both in terms of architecture as well as construction. Most of these buildings are now in very poor condition and—if not vacant—they mostly host low-profile and very basic retail units. Very significant is the increase in population resulting from the intensive flows of foreign immigrants some of whom have encouraged the reuse of existing Islamic religious sites.

Along with the intervention of the municipal offices and the new Orthodox Cathedral, a major intervention is Eleftheria Square, designed by Zaha Hadid.

In addition, many historic buildings built with weaker materials such as adobe bricks have collapsed over the past two wet winters leading to a major demolition campaign of potentially dangerous structures. This has left large empty frightening gaps in the Greek Cypriot side of the city. These are mainly used as car parks to meet the high demand for more cars in the old town.

On the north side, there are several developments in the historic center that are worth exploring. The great bazar has been completely restored and now boasts very large spaces, with the stalls partly used as fruit and vegetable shops, some antiques, and a second-hand bookshop. While the Agia Sophia and the Bedestan (the former Orthodox cathedral dedicated to Saint Nicholas) are currently being renovated, many houses and small mansions have been increasingly refurbished, and the overall consistency of the original and ancient building stock is remarkable. The greater coherence of this older heritage may mainly be due to the fact that in the south side, during the financially prosperous 1950 and 1960s, a lot of demolition and replacement took place while in the north the pace of development and of business was much slower resulting in the preservation of the old built heritage and an urban landscape with more coherence and old flair.

However, despite the intense memories and the strong interest we may have for what we consider to be the heart of the island, Andreas notes that the general feeling shared by most of the Cypriot friends and acquaintances is that the divided city center suffers from an underestimation and lack of interest by most of the population (south and north side): they definitely have other housing preferences and options, also in terms of lifestyle. Apart from the conservation and historical preservation of monuments, the economic and cultural interest of Cypriots in Old Nicosia seems to be very low, and this is a perspective challenge, both for the preservation and revitalization of the historic city, and for the intensive land consumption and overexploitation of other areas on the island affected by intensive development.

References

Abu-Orf H (2005) Collaborative planning in practice: The Nicosia Master Plan. Planning Practice and Research 20(1): 41-58

Bricocoli M, Savoldi P (2012) Il disastro pianificato e l'innovazione silenziosa. Due (quasi) progetti urbani a contrasto. Archivio di Studi Urbani e Regionali 98:74-90

Bricocoli M, Breckner I (2012) 'Geschlossene Irrenanstalten' als Ressourcen der Stadterneuerung / 'Prisons of Madness' as Sources of Urban Innovation. In: Eisinger A, Brodowski N, Seifert J (eds) Urban Reset. How to Activate Immanent Potentials of Urban Spaces / Freilegen immanenter Potenziale städtischer Räume. Birkhauser, Zurich, p 146–155

Bricocoli M, Curci F, Savoldi P (2015) A cose fatte: tre progetti urbani a Milano, Copenhagen, Amburgo. Urbanistica 153, p 67-85

Damgacı AY, Dağlı UU (2018) Shifting Boundaries of Divided City Nicosia through Social Actors. Space and Culture 21(4):482-494

Ewers CM (2018) The Nicosia master plan: historic preservation as urban regeneration. University of Oregon Press, Corvallis

Gaeta L (2018) La civiltà dei confini. Pratiche quotidiane e forme di cittadinanza. Carocci, Roma

The Experience of the Nicosia Masterplan. A Conversation with Ali Güralp and Agni Petridou

Ali Güralp, Agni Petridou, Alice Buoli, and Oana Cristina Ţiganea

Abstract This contribution presents a conversation with Agni Petridou and Ali Güralp, leaders of the Nicosia Masterplan bi-communal team, implementing common urban planning and heritage preservation initiatives between North and South Nicosia since the 1980s. The outcomes of this discussion reveal a profound and enduring professional as well as personal commitment to the revitalization and strategic future development of the city, manifesting in a life-long friendship that transcends the political and physical divisions.

1 Defining Territorial Fragility in Cyprus Through Nicosia Masterplan's Lens

Alice Buoli (AB) and Oana Cristina Ţiganea (OCT)[1]

How would you define the issue of "territorial fragility" in the context of Nicosia and how have such conditions changed since the early 1980s when the Nicosia Masterplan (NMP) was first started? How has the NMP recognized and contributed to dealing

[1] This conversation was conducted and recorded in July 2022. While the initial set of questions prepared by the editors served as a general framework and prompts for the conversation, they were adapted to the responses and reactions of the participants. The recording has been transcribed with the support of Neofytos Christou, an architecture student at Politecnico di Milano, and then edited by Buoli and Ţiganea.

For a more detailed and further discussion on the Nicosia Masterplan, please also see the essays by Bricocoli and Chrysochou, and Gaeta and Pasqui in this volume.

A. Güralp · A. Petridou
Nicosia, Cyprus

A. Buoli (✉) · O. C. Ţiganea
Department of Architecture and Urban Studies, Politecnico di Milano, Milan, Italy
e-mail: alice.buoli@polimi.it

O. C. Ţiganea
e-mail: oanacristina.tiganea@polimi.it

© The Author(s), under exclusive license to Springer Nature Switzerland AG 2023

A. Buoli and O. C. Ţiganea (eds.), *Territorial Fragilities in Cyprus*, Research for Development, https://doi.org/10.1007/978-3-031-36076-3_3

with different kinds of (social, environmental and material) vulnerabilities of the city?

Ali Güralp[2] (AG)

The Nicosia Masterplan was initiated during a period of great fragility in Nicosia. Politically, socially, and economically, the atmosphere was tense, and people found it difficult to meet and discuss matters. Despite these challenges, the mayors of both sides recognized the need for action and began to address the issues in a fragile environment.

In the early 1980s, the mayors began meeting at the Ledra Palace Hotel, seeking to break the ice and find common ground. This led to the development of the Nicosia Masterplan, which focused on town planning and infrastructure improvements designed to address the city's fragilities and bring the two communities together. Through mutual understanding and the solution of shared needs, the plan was able to make progress towards this goal.

Prior to the Nicosia Masterplan, development at the community level had been limited due to the separation of the two communities. However, progress was made at the municipal level, with twin projects and developments on both sides of the city aimed at preventing the differences between the two communities from growing. The success of these efforts was due in part to the meetings held at common ground locations like the Ledra Palace Hotel and the airport area.

The opening of checkpoints in 2003 marked a turning point in the development of NMP, allowing for faster progress and the gradual softening of fragilities. While progress had been made prior to this point, the increased mobility and interaction between the two communities enabled by the opening of checkpoints helped accelerate this process.

Agni Petridou[3] (AP)

I agree with Ali when he says that our experience with the Nicosia Masterplan was unlike any other town planning process. It was not just a technical collaboration; we had to establish new mechanisms and build trust among our team members because it was a unique initiative in Cyprus. As Ali mentioned, the local leaders—Mustafa Akıncı[4] and Lellos Demetriades[5]—were very supportive and effective in resolving political issues. This made our work much easier. Our primary objective was to address the infrastructure problems that arose due to the city's physical devastation

[2] Ali Güralp is a Civil Engineer and served, between 1994 and 2021, as the team leader of the Nicosia Masterplan North, Head of the Department for Projects. Since 1994 he has been working on bi-communal project cooperation and implementation.

[3] Agni Petridou is the Director of the Technical Services of the Nicosia Municipality and former Head of the Nicosia Master Plan team (Greek Cypriot part). She is also a member of the National Committee of the International Council on Monuments and Sites ICOMOS and of the Team of Experts of the Priority Action Program of UNEP.

[4] Mayor of North Nicosia Municipality from 1976 to 1990. Later he became the president of the Turkish Republic of Northern Cyprus from 2015 to 2020.

[5] Mayor of Southern Nicosia Municipality from 1971 to 2001.

caused by the partition. We had to create a common infrastructure and prepare a plan for the city's future, taking into account the existence of the Buffer Zone and the possibility of reunification. Each proposal had two phases, one for the current situation and another for a unified city. For example, our traffic and transportation strategy included solutions for both phases. This experience was very special, and I consider it the most important period of our career. Working in such a unique situation helped me form strong bonds of friendship with my team members, like Ali, who worked with me since the initial stages of the project.

OCT

Could you provide an update on the current status of the Nicosia Masterplan? Is the project still being implemented with a focus on both the present and potential future scenarios in the event of a reunification? Furthermore, does the project team still maintain the belief that the city may one day be reunified, and as such, the planning strategy should reflect this possibility?

AP

We have always worked together with regard to the town planning strategy. For instance, during the second stage, after 1981, we began anew in the early 2000s with a new planning strategy, and currently, we are working on the two Area Schemes of Nicosia: one for the Greek Cypriot community and the other for the Turkish Cypriot community. The same planning team develops both Area Schemes based on the same principles: it is important to prepare the city for a potential reunification. However, this is a political issue that we cannot guarantee, but as technical experts and town planners, we do our best within our capacity. Since we have different funding sources now, the coordination is not as smooth as before. Prior to joining the European Union, we worked with the same funding. We had the funds allocated to the two communities through the UNDP, from the European Union and from USAID, which were coordinated to submit common proposals for funding, ensuring synergies among our projects. After entering the European Union, the Greek Cypriot community has access to structural funds while the Turkish Cypriot community is funded by the task force of Europe, leading to a looser coordination between the two communities. Nevertheless, we still strive to be coordinated and maintain a common strategy through the Area Schemes that are detailed urban plans, such as the "Piani Particolareggiati" in Italy.

AG

Yes, if I may add, we as technical experts have managed to establish a solid foundation. The Nicosia Masterplan report served as the basis for the second phase, which was then reinforced by a new vision that outlined plans for both sides of Nicosia. Our aim was to ensure that development needs on both sides were aligned, and we achieved this through joint planning efforts. The Area Scheme in the South and the recently completed Strategic Plan further strengthened our unified approach. The planning tools and books we have produced will guide future planners and municipal personnel towards a unified Nicosia. However, as technical professionals, we cannot control political developments, and our priorities may differ from those of politicians. As funding sources have been separated since 2004, it has become more challenging to implement joint projects in a timely and coordinated manner. Although we managed to implement some projects on both sides of the city, they were not always implemented in parallel, leading to disparities in development. Unfortunately, the northern side, which is not receiving funds as easily as the southern side, is lagging behind. The current political situation is not helping, and it seems unlikely that the situation will improve anytime soon.

OCT

Well, the current political situation worldwide is not conducive to reconciliation in any part of the world. We were actually hoping that you could shed some light on the impact of time and how different approaches have evolved over time. When you mentioned how you built a strong technical foundation, including the Nicosia Masterplan reports and the Area Schemes, we were curious about the approach of the younger generation who have grown up in a divided city. Their perspective on the vision for a unified Nicosia may be different from yours, so we wonder how they will be motivated to continue the work that you have started?

AP

It does not depend on the individuals in technical positions, but rather on who establishes the vision. If the mayors are individuals who genuinely believe in a shared future, they set the vision, and plans must follow accordingly. This is a crucial matter that cannot be left to the discretion of each individual planner. It requires political will to maintain the city's prospect for reunification. Fortunately, in our case, the new Area Scheme, which is the current planning strategy, was prepared by the same team. For the Greek community, we conducted a global tender to select the best planning team to develop the project. It was not only prepared by Cypriots but also an international team comprising British and Italian planners. When Ali conducted the tender, and the funding came from the World Bank, the same team won the tender in the other community. As a result, both communities' Area Schemes were prepared by the same group of experts. The current mayors,[6] Mehmet Harmancı and Costantinos Yorkatzis, believed in this vision and collaborated to make it a reality. During the public consultation process for the Area Scheme, the Nicosia Masterplan was invited as a significant stakeholder to express its experience and desires. It is important for political leaders at the local level to continue to promote the concept of a shared future.

AG

I completely agree with Agni's statement that collaboration between planners depends on the political will of the mayors. I personally know both planners from the municipal personnel on both sides, and they are willing to work together and implement the work we produced jointly. The technical teams are collaborative and want

[6] At the time when the interview was recorded (July 2022).

to work together, but politicians, especially mayors, play a crucial role in promoting collaboration. So far, the mayors have been positive in this regard. However, I would like to emphasize the role of the European Union in facilitating collaboration between the two sides. The EU should take a more active role in encouraging collaboration and funding projects outlined in the Area Scheme or Strategic Plan. The EU can serve as a valuable facilitator for implementing the plans, given the shared mindset and mutual desire for implementation among the involved agencies. If there is no push to collaborate, the needs of both sides will diverge over time, and each side will try to resolve their problems independently. Therefore, it is essential to encourage and support the implementation of the Area Scheme and Strategic Plan on both sides with the crucial financial and institutional support of the EU.

AP

Yes, Ali is absolutely right. He is correct in pointing out that there are currently two separate funds with no intermediary. The European Union needs to promote bi-communality through projects, such as those identified by the Nicosia Masterplan. There have been many proposed bi-communal projects for the city, but the most important and key project could be the one focused on the Buffer Zone, which is crucial from an architectural [and heritage preservation] perspective. There is a project that has already been carried out together, and it could serve as a basis for finding support for these buildings as a means of bringing the two communities together. However, it is crucial that the funding for this project remains distinct from the funds allocated to each community, because when funds are announced, all of the ministries and municipalities in Cyprus rush to claim the money. The same situation occurs with the European Union's funding for the North. As Ali points out, separate funding for bi-communal projects is necessary to keep collaboration alive. Unfortunately, such funding does not yet exist.

OCT

You actually anticipated our next question which is: what is the current approach regarding the Green Line, with a particular emphasis on heritage buildings and sites, as well as the issue of property disputes? During our visit to Nicosia, we observed that there are still unresolved issues with individuals who have one wall of their property located in the Buffer Zone, rendering the property unusable. We also witnessed numerous instances where properties were demolished because they were in close proximity to the Buffer Zone. Is there a plan in place to address these challenges, or is it difficult to intervene due to the various sources of funding?

AG

As Agni pointed out, the Buffer Zone is the most common ground for development. However, due to the political situation, it has been challenging to undertake actual physical work in the Buffer Zone. Therefore, the strategy has shifted towards developing the area just outside of it. By upgrading these areas, particularly in the North, which is currently one of the most disadvantaged areas of the city, we can transform the economic and social aspects close to the Buffer Zone. This could create pressure on politicians to open up the area for development.

While continuing to push for implementation in the Buffer Zone, we need to focus on upgrading the areas adjacent to it. This will not only improve the city but also encourage political action. Moreover, there is a project funded by the EU that Agni and I will be part of, which is the feasibility study for improving the social, physical, and infrastructural aspects of the walled city in the North. This study will guide us towards making improvements that will benefit both the northern and southern sides of Nicosia.

2 Building Common Knowledge on Nicosia and the Green Line

AB

Considering what you just mentioned, we were wondering, on a technical level, how did you produce a common cartography, spatial analysis, and surveys of both cities, as well as the Green Line? We know that you tried to perform a survey of the facades of Green Line buildings, but we were curious to know more about how you worked together in practice to share knowledge on the physical aspects of the city.

AP

During the preparation of the Area Schemes, we conducted detailed surveys of the existing conditions on both sides. For this, we engaged a team of consultants to carry out the work while we, as members of the steering committee, oversaw the project. In terms of the Buffer Zone, we formed common teams consisting of four Turkish Cypriots and four Greek Cypriot architects, engineers, and topographers from the Nicosia Masterplan. We spent three years working in the Buffer Zone, surveying and redrawing the facades of 365 buildings and surveying the infrastructures. Due to safety concerns, we were unable to enter the buildings, which had collapsing roofs and were infested with mice. Our goal now is to conserve and protect these buildings, provide them with necessary repairs to prevent further collapse, and prevent their demolition. Our surveys are ongoing, and we plan to publish the results as we believe they are valuable and informative.

OCT

We were wondering about the legal powers of the Nicosia Masterplan on both sides of the Green Line, as there seem to be different funding sources and regulations in place. Can you explain the legal instruments that are used to transform the planning project into regulatory norms for building and development? We understand that this may involve political issues as well.

AP

In the North, the equivalent of the Area Scheme is the Strategic Plan, which has been prepared but has not yet been legalized. Planning laws are produced by the town planning department under the Ministry of Interior. As the municipality or Nicosia Masterplan, we prepare the Area Scheme or Strategic Plan and hand it over to the town planning department. However, the technical personnel in the town planning department are responsible for implementing the plan but do not have the legal tools to enforce it. Therefore, the law must be passed, and we ensure that the implementation goes according to the plan.

There is a challenge in making the plan mandatory after it has been completed. We use the same planning law that is based on the British planning law at a certain level, and we follow two processes that are not completely different between the two communities. In the Greek Cypriot community, we went through the process to make the plan mandatory and it is now legally binding. However, the process takes a long time due to bureaucracy. I am not sure how long it takes in Italy, but in our case, the group of experts prepared the Area Scheme in one year, and it took us three years to make it mandatory. Nonetheless, it is now part of the law. Ali, do you have any information on how long it takes in Northern Cyprus?

AG

The situation is more challenging here as the government changes frequently. For instance, when we prepared the Strategic Plan, the interior minister was very supportive of making it mandatory. Unfortunately, before we presented it to her, the government collapsed, and the new government is not as supportive of the plan. Thus, the plan is still waiting for approval. As I mentioned earlier, there must be projects or plans that create a necessity for politicians to consider these matters. If there is no progress in implementing planning laws or matters, no one will take action. However, if the public demands improvements and the need for ready laws become apparent, politicians will have to pass the law to meet the people's needs. In this regard, I believe the EU plays a crucial role in supporting the implementation of projects within the Area Scheme, the new vision, or the Nicosia Masterplan.

3 Current and Future Challenges for the NMP and Its Planners

AB

I think you touched upon very crucial problems that can push politicians to take action. The Nicosia Masterplan has already addressed some of these issues, such as the preservation of heritage sites within the Buffer Zone. However, do you think that there are other urgent matters that can push politicians to act? For example, has the Nicosia Masterplan addressed or considered issues such as the impact of climate change or earthquake risks, along with the preservation of heritage sites within the Buffer Zone?

OCT

Additionally, do you think that the energy transition could be a major topic of discussion for the entire island, and not just limited to Nicosia?

AP

The issues you mentioned are broader and fall under the purview of the bi-communal technical committees appointed by the two communities to facilitate discussions on solving the Cyprus problem at a national level. These committees focus on specific areas such as environment, cultural heritage, and other important issues. The technical committee for cultural heritage, in particular, has been very successful in restoring religious monuments like churches and mosques. The Nicosia Masterplan, on the other hand, primarily focuses on Nicosia and the city center, which is of great importance as the capital city and for the symbolic meaning of the Buffer Zone. As a result, the Nicosia Masterplan mainly deals with technical issues related to the Buffer Zone, as it cannot resolve the political issues surrounding it.

OCT

Besides the Green Line, what other needs do you foresee needing to be solved in Nicosia? You mentioned a new feasibility project in the North.

AP

There are infrastructure improvements needed for both parts of the city, such as the storm water system. There are also other difficult issues, such as the river that runs through both parts of the city, which could be turned into a linear park. The Nicosia Masterplan team proposes several solutions every year, but unfortunately, they haven't been funded yet. It's not just a matter of funding though, there are also political difficulties that need to be resolved in order for us to collaborate and work together. However, external entities such as the EU or UN could facilitate this process, as they have done in the past.

AG

I would like to add that cultural heritage preservation in the walled city is crucial for both sides. It not only enhances the physical development of the area, but also attracts the original owners to make investments in their properties. The education sector could serve as a facilitator for this. If we manage to bring back some faculties of universities to the walled city, students will need accommodations, which will in turn require the restoration or upgrading of many of the currently dilapidated buildings.

AP

For example, in the South, one of the oldest and most historical schools, the Faneromeni, was donated by the owners, the Church, to become the School of Architecture of the University of Cyprus. Additionally, the municipality bought a lot of properties, restored them, and turned them into student hostels. If the North could do the same and promote common functions, they could be implemented in the Buffer Zone. Architecture schools are pioneers in these initiatives and could collaborate well with us.

AG

As I mentioned earlier, the focus should not just be on the Buffer Zone, but also on attracting the original land and building owners in the walled city to come back and invest. The presence of students who require accommodation, food, and entertainment can provide an incentive for these individuals to invest in the walled city, leading to more rehabilitation activities. Education, cultural buildings, and tourism are sectors that could facilitate improvements and draw the attention of politicians to take action in terms of funding and legal frameworks. However, the North does not have the funding to support these initiatives on its own, and collaboration with the municipality is necessary.

For instance, the Ledra Street and Kyrenia Avenue economic activity spine is an area that needs improvement, as only Ledra Street in the South has been developed and pedestrianized. The EU could potentially fund the pedestrianization of Kyrenia Avenue, connecting it to Ledra Street as a continuous commercial axis, and encouraging collaboration among the shopkeepers and landowners on both sides. In order to initiate these efforts, the EU could be the sole facilitator and highlight the need for politicians to take action.

AP

The reality is that we need an international or European institution to support us on our path to Europe. The interests of other (Cypriot) municipalities are completely different from ours, as they do not face the same challenges that we do.

AG

I agree with Agni. The Northern part being outside the EU regulations—since the *acquis communautaire* is suspended—is not necessarily obliged to abide by them, thus the priorities are not easily acknowledged or appreciated since there is no precondition in the existing system to fulfill in order to proceed to the next stage. Hence the goodwill falls short and the organization is not understood. So, the two sides speak two different languages and they don't see the local perspective. However, we, along with the people in the two municipalities, have over thirty years of experience in public works. We need a mediator who can communicate and align the requirements of both sides and facilitate cooperation.

OCT

This leads us to another question we have, which is about finding a common ground, not only in terms of physical space but also as an aid for negotiation. I agree that an external body is needed to facilitate this process, as it should be neutral and not take sides. We were thinking if the example of the Home for Cooperation, which is inside the Buffer Zone and an exception to everything that happens in there, could be applied in other crossing points of the city. This could be done through bi-communal projects that aim to open up checkpoints and implement a strategy for cooperation. Do you think this is feasible?

AP

We proposed this, how many years ago? Was it four or five years ago? We worked together with Ali to identify positions, prepare the letter, and send it to the mayors. However, nothing happened because the issue of crossings is a political one and is difficult to resolve.

AG

We pushed as much as we could up to the point where there were political decisions to make. We even tried to push the politicians, but their agenda is different. But what I am suggesting now is that, of course, as a location, the Home for Cooperation is a good place to do bi-communal projects, seminars, or conferences. However, what we need is an organization from the EU to come and facilitate these collaborations or promote these good practices necessary to bring the two communities together.

So, we are here to identify the projects that will facilitate the coming together of the two communities. If such an organization is formed from the EU, I think everyone will take an interest in this, and both sides will benefit. But, as I said, we have the plans and tools ready…

AP

… and the ideas…

AG

…the ideas and everything. So, if such a bi-communal organization is established, composed of local people like us and representatives from the EU, whose work is focused on promoting bi-communal projects, it could be specific to the Nicosia's Walls project, for example. There are numerous topics that can be addressed together, and both sides would benefit from this. Politicians will not be able to reject this because it would aid their cases and help them with re-election.

AP

Recently, we had visits from high-level European officials like Margaritīs Schinas, and last week, the Commissioner for Development, Elisa Ferreira, who allocates structural funds, was here. During their visits to Eleftheria Square, I asked both officials about the Buffer Zone, and they both said they would be very supportive of funding a project like that. The problem is that we need a facilitator to convince the leaders of the two sides and the European Union commissioners to act as our mediator and help us promote these ideas, as Ali suggested.

AG

In the South, they are lucky to have people like Agni, and the leaders listen to them. However, in the North, we need to be recommended by facilitators because technical people are not considered competent enough to be heard regarding development. Therefore, we will require support from these facilitators. For instance, they might request people who were part of the Nicosia Masterplan earlier on and are willing to contribute to assist them. Unfortunately, we are all aging now, and the old Nicosia master planners are retiring or moving away from the idea of bi-communality, given the contemporary political situation and changing population needs. This is perhaps the last chance where we still have some people like us who are willing to contribute. If this momentum does not continue in the coming years, these ideas will eventually fade away, and there will be no one left to support them.

OCT

What about the effectiveness of communication and collaboration? You mentioned that there have been significant structural, social, and cultural changes over the past 40 years, and that the Nicosia Masterplan has evolved and become a reference point that people listen to. But what about the current state of critical mass communities in both North and South Nicosia? Will they be more willing to push for common projects and support your efforts?

While an external body can facilitate political discussions, it's also important to consider the feasibility of creating critical mass communities in both regions.

AG

The community is currently in a precarious position and reluctant to take risks due to the poor economic conditions in the North, caused by the decline of the Turkish Lira. People are afraid to invest because they fear losing their livelihoods, which are mostly denominated in Turkish Lira, and high inflation rates only exacerbate the situation. However, the younger generation is eager to invest in the wall(ed) city and start-ups are looking for opportunities to make an impact. Although funding from parents can only go so far, if there were an external facilitator like the EU, young entrepreneurs could be supported through initiatives like the Innovative Entrepreneurship Hub or CYENS Center of Excellence, with actual concrete funding. If the European Union supports such initiatives, especially those involving cultural heritage and bi-communal projects, the younger generation, who represent the future, is ready to create change. For instance, I have two children who are willing to invest in our building located in the walled city, but we cannot rely on the Turkish Lira or Turkish support. By having the EU as a partner, we would have a clearer vision for the future and a better chance of success.

AP

This CYENS project is of great importance, as it is situated close to the Buffer Zone and involves collaboration between the municipality, the University of Cyprus, and the University of London. A similar project can be initiated on the other side of the Buffer Zone, as Ali suggested. The young entrepreneurs are very enthusiastic about it, and it is a brilliant idea that can be implemented at a bi-communal level. There is no lack of ideas, and many initiatives are already in progress. However, there is a need for bi-communal funding and a mediator to support these efforts. Ali, do you have any thoughts on this matter?

AG

Yes, there is an innovative entrepreneurship hub in the North, which is a European Union project, but it seems that the Nicosia Masterplan ideas are not being fully utilized, despite their importance for the walled city. However, there is a collaboration with CYENS, so this facility provides a good opportunity for both sides. Especially for the younger generation, it can be used to implement Strategic Plans, Area Schemes, and upcoming projects that will emerge from feasibility studies currently being conducted.

AB

One last question: you mentioned in the beginning that the time spent on the project was very important in your life and professional career. We would like to ask you what you see as the future of the Nicosia Masterplan outside of Nicosia? Do you see it as a model that could be extended to other parts of the island? Finally, we would like to hear your personal thoughts about the future.

AP

Certainly, the Masterplan model is of great importance and has the potential to be extended to other areas experiencing conflict. From the perspective of those involved in the Nicosia Masterplan since its inception, it is clear that collaboration and the development of personal relationships have been essential to the success of the project. Moving forward, we aim to support and empower current members of the Masterplan while continuing to foster cooperation through bi-communal collaborations. For instance, Ali and I plan to work on a project together, drawing upon our experience in navigating the unique challenges of bi-communal work. We believe that our experiences can be transferred to others seeking to engage in similar endeavors.

AG

We have established a close-knit relationship over the years, starting from the early stages of our professional careers and expanding into personal friendships. This bond has grown stronger over time, as we have seen each other through important milestones such as our children's weddings. Our friendship is a testament to the power of collaboration and the importance of building relationships between communities, regardless of nationality or background.

The success of the Nicosia Masterplan is not limited to its impact on the city's physical landscape, but also its ability to bring people together from all over the world. Our team has gained friends and supporters from various countries such as South Africa, Germany, and Italy, who have also embraced the vision of a united and peaceful Cyprus. We believe that our experiences can serve as an example for others, whether they are in conflict zones or not.

Moving forward, we are excited to continue our work on the walled city in the North through a feasibility study project that is soon to be signed. We remain committed to our mission and are grateful for the opportunity to continue working together with our friends and colleagues, with the hope that our collaborative efforts will endure for as long as we live.

Fragile Tangible and Intangible Constellations in a Divided City. Urban Planning and Architectural Preservation Perspectives in Nicosia

Alice Buoli and Oana Cristina Ţiganea

Abstract The chapter discusses and unfolds the dimensions of territorial fragility in Nicosia in light of Cyprus' partition and its multiple reverberations on the city's-built environment and everyday urban life. The essay builds on academic debates on contested spaces and urban division from a planning and heritage preservation perspective, highlighting the most prominent lines of research and interrogating the same definition of Nicosia as a "divided city". The chapter also recounts the genealogy of the Green Line and the Buffer Zone, from tracing a line on a map to its implications on the ground and on people's perceptions and identities. The second part of the essay draws on and recounts on-site and empirical work performed in the city. Through collective "walks" on both sides of the Buffer Zone—adopted as an explorative method and a narrative device—a series of "constellations" of open and built spaces, semi-public and secluded lands, images and narratives emerge. Even though discontinuous and episodic, these spatial narratives provide a composite, multi-dimensional, and multi-layered image of Nicosia and its tangible and intangible urban conditions along and inside the Buffer Zone. These narratives allowed a horizontal *(in space)* and vertical *(in time)* reading of Nicosia's cityscapes. Ultimately, the chapter provides the thematic and methodological elements of the research-by-design explorations proposed in this volume, drawing on and feeding back the theoretical and empirical contributions of the book.

1 Introduction

Upon arriving in Nicosia from the south and driving along the highway that connects the international airport of Larnaca to the capital of Cyprus, one may notice a dense sequence of barbed wire fences and signs indicating restricted access to lands and

A. Buoli (✉) · O. C. Ţiganea
Department of Architecture and Urban Studies, Politecnico di Milano, Milan, Italy
e-mail: alice.buoli@polimi.it

O. C. Ţiganea
e-mail: oanacristina.tiganea@polimi.it

© The Author(s), under exclusive license to Springer Nature Switzerland AG 2023
A. Buoli and O. C. Ţiganea (eds.), *Territorial Fragilities in Cyprus*, Research for Development, https://doi.org/10.1007/978-3-031-36076-3_4

fields on the two sides of the road. Despite being a few kilometers away from the UN-managed militarized Buffer Zone that divides the island since 1974, these visual markers stand against the Cypriot dry and hilly landscapes along the highway, underscoring the seriousness of trespassing.

The initial perception of Nicosia's partition already arises as one approaches the city's suburbs, where most of the Greek Cypriot residents have established themselves over the past fifty years, in an effort to create a distance from the continuous tangible reminder of the partition and its everyday life emotional and political implications.

Getting closer to the walled city, high-rise residential buildings mark the skyline, silhouetted against the Pentadaktylos / Beşparmak (literally "five fingers") mountain that bound the city north.

When viewed from the south, the mountain serves as an additional visual presence of the "other side" through the display of flags of the Turkish Republic of Northern Cyprus (TRNC) and Turkey, which are reproduced as "out of scale" signs in the landscape, visible even at night.

Arriving at the (southern) city center with the first light of the evening, the recently completed Eleftheria Square by Zaha Hadid Architects—a contemporary intervention juxtaposed on the sixteenth-century Venetian Walls—glows as an "out of the ordinary" element in the old town cityscape.

Passing west from Paphos Gate and crossing into the TRNC at the Ledra Palace checkpoint, that opened in 2003 for the first time in 30 years, the moat opens the view to a recent and shiny green football pitch. This meaningfully mirrors the Orphea football club stadium on the eastern side of the city, close to Famagusta Gate.

On the north as well as in the south, each residential and public building's rooftop has its own water tank, indicating the punctuality of the water infrastructure in the area. In 2021, during one of the hottest summers of the past century in the Mediterranean, the water tanks highlighted one of the common challenges that the island and the city are currently facing, making water a topic crossing everyday life as well as geopolitical relations in the island.[1]

Leaving Nicosia to Famagusta on the eastern side of the island, the cityscape is dominated by apartment buildings, with a handful of notable landmarks breaking up the suburban landscape: the minarets of a series of newly built big-scale mosques and a public "parade" site along Doktor Fazıl Küçük Bulvarı. Along the highway, the summer dry and yellowish landscape is occasionally interrupted by quarries, industrial areas and small farms.

This synthetic and partial narrative of "landing" and "departing" in/from Nicosia accounts for a city that is closely shaped by the trauma of the conflict and the events that occurred since the early 1960s on the island. Both urban settings reflect in different ways the long-term effects of the partition together with the narratives that have been developed around the Buffer Zone, perceived as the physical manifestation of the "Cyprus problem". Such perceptions, expectations, and disillusions are inscribed in the city's stones, further echoing the stories and voices of its inhabitants.

[1] As discussed in the Chapter "Civic Water. Bridging Culture, Nature and People".

In the following pages, we will explore the role of the partition in the production of Nicosia's cityscapes and built environment, with a particular focus on the urban dimension and its tangible and intangible heritage elements.

The Buffer Zone and the spaces in its proximity will be presented in their current material configuration, considering the past events that led to the translation of a line—the Green Line—from a map into its material configuration on the ground. A line that broke into people's lives and daily habits, family links and neighborhoods, properties, and even buildings. Such an overview aims to disclose not only the city's past and current fragilities, but also the role of the Green Line in conveying potential imaginaries and visions of reconciliation among the local communities on both sides.

The following paragraph will first introduce the academic debates on contested spaces and urban division from a planning and heritage preservation perspective, highlighting the most prominent lines of research as well as the main criticisms on the expression "divided city".

The chapter unfolds, then, the main dimensions of territorial fragility in Nicosia, building on the empirical work performed during a collective study mission in Cyprus in the summer of 2021[2] and more recently in 2023 through a research visit.[3]

In the following section, the chapter proposes a series of narratives understood by the authors as "constellations" *à-la* Walter Benjamin,[4] of open and built public, semi-public and secluded spaces, emerged thanks to a series of collective "walks" on both sides of the Buffer Zone, adopted as an explorative and analytical method, and a narrative device. Even though discontinuous and episodic, these narratives develop a composite and multi-layered image of Nicosia, of its tangible and intangible urban conditions along and across the Buffer Zone, allowing both a horizontal *(in space)* and vertical *(in time)* reading of Nicosia's borderscapes (Fig. 1).

2 Bridging Urban Planning and Heritage Preservation Debates on Contested Spaces

A general understanding of "divided cities" as suggested by Carabelli (2018: 7), describes them as places where ethnic, social and political segregation occurs and materializes through the construction of physical barriers, such as buffer zones, fences, walls and internal checkpoints to cross them, that limit mobility between inside and outside or between one side and the other, most of the time related to forms of (violent) conflict and confrontation. In fact, urban divisions usually reflect conflicts and contestation at higher territorial levels, such as in the case of Bosnia and Herzegovina, Israel/Palestine, and Lebanon (*ibidem*: 8).

---

[2] Please, see the introductory chapter by the book editors for more details.

[3] Alice Buoli was hosted as visiting scholar at the Department of Architecture of the University of Cyprus (UCY) between March and April 2023 with the support of prof. Socrates Stratis.

[4] Gilloch, G., 2013. Walter Benjamin: Critical Constellations. John Wiley & Sons.

Fig. 1 The Buffer Zone seen from Kaimakli neighborhood, just outside the walled city, South Nicosia. In the background the Pentadaktylos/Beşparmak mountain. Photo by Oana Cristina Ţiganea, 2021

Looking at the scientific production on divided cities over the past decade—specifically related to the field of urban studies, architecture, and heritage preservation—one can encounter a variety of research products that have a clear "comparative" and "case-studies" character, that provide an array of conceptual and operative approaches to contested urban environments. The most cited and prominent works in the Anglo-Saxon realm, published in the past 10 years on these topics include an ample collection of books, out of which stand out the following works: "Divided Cities. Belfast, Beirut, Jerusalem, Mostar, and Nicosia" by Calame and Charlesworth (2011), "Planning in Divided Cities" by Gaffikin and Morrissey (2011); "City and Soul in Divided Societies" by Bollens (2012); "The Divided City and the Grassroots The (Un)making of Ethnic Divisions in Mostar" by Giulia Carabelli (2018); "Urban Heritage in Divided Cities: Contested Pasts" by Ristic and Frank (2019); "Bordered Cities and Divided Societies Humanistic Essays of Conflict, Violence, and Healing" by Bollens (2021).

The book by Calame and Charlesworth (2011) constitutes an attempt to identify patterns and common features of five different "divided cities" (Belfast, Beirut, Jerusalem, Mostar, and Nicosia), starting from the assumption that "developmental patterns illustrative of a generic divided city phenomenon might emerge" from such a comparative approach (*ibidem:* xi). The book's premises are grounded in a reconstruction planning workshop that the authors conducted in Mostar (Bosnia), in the years that followed the end of the Balkan wars. Considering the disciplinary perspective of the authors, which combines architecture, urban design and planning, and preservation, their aim is not to provide "solutions" (plans or policies). Instead, the book intends to provide evidence and knowledge on the pre-partition causes and conditions for each city, identifying common features across the different contexts, that might guide a more effective management of the urban division (*ibidem*: x).

On-site research in each of the five cities was conducted through direct observations and interviews with local stakeholders and citizens, to respond to a series of basic questions related to the genealogy of the partition and its manifestation in time; the social and physical impacts of the division in the cities' everyday life and of the local populations; issues related to land, property rights, collective security and sovereignty. An overarching observation appears quite relevant to be mentioned: spatial markers of division (walls, flags, fences, no man's land) despite their initial "temporariness", have become, in many cases, permanent elements of the urban environment "forcing divided residents to grapple with life under siege" (*ibidem*: 1), thus perpetuating the idea of a never-ending conflict.

As suggested by Silver (2010) one of the main contributions of this volume regards the recognition that the process of urban division is not a sudden event, but instead "a gradual, predictable, and avoidable sequence in which insecurity gives way to ethnic violence and internal partition. First, ethnic identity is politicized and serves as the basis of social organization (…) Second, insecurity and conflict are catalysts for the clustering of threatened communities who 'stick to their own kind' for protection. (…) Third, enclaves assume political and symbolic significance, forcing residents to sacrifice for a larger cause. Fourth, the spatial boundaries between ethnically homogenous, politicized enclaves are drawn along dormant physical fault lines that become concrete and impermeable. Interfaces where ethnicities once mingled become dangerous no-go zones policed by the combatants and set off with markers. Public spaces become sites of greater, not less sectarian violence. Barricades become walls. Finally, states or external forces consolidate and institutionalize the boundaries in an effort to reduce bloodshed" (2010: 249).[5]

From a different perspective, the underlying argument of "Planning in Divided Cities" by Gaffikin and Morrissey (2011) is the recognition of the role that urban planning and policies play in "the shaping of social space" which in turn is grounded "at the heart of many urban conflicts, particularly those that connect issues of territory and identity" (*ibidem*: 7). Seen from this perspective, urban planning and policy cannot be treated independently from the resolution of urban conflicts and thus have to prioritize issues of capacity-building for local communities.

In this sense all cities can be considered divided "in that their 'publics' and stakeholders have differential access and vested interest, marked by distinctions of class, ethnicity, and gender" (*ibidem:* 9). For the authors contested urban spaces can be

---

[5] As we will see in the following paragraph, such process describes quite literally the different stages of Cyprus and Nicosia's partition in time.

looked at through a dual lens: that of *pluralism*, that is, relating to imbalances of power, wealth and status between distinct social groups (an example are forms of ethnic or "racial" division/segregation in North American cities); and that of *sovereignty*, where pluralist disputes over equity and access to urban space are intertwined with ethno-nationalist conflicts over the legitimacy of the state itself (as in the case of Belfast) (*ibidem*). Here the term "divided" is used in a quite loose meaning, not only to refer to the group of cities identified by Calame and Charlesworth but broadening the gaze to cities that witness various kinds of division and contestation. In fact, the authors acknowledge the ambiguity and ubiquity of the term "divided city", which thus remains elusive while staying open to various definitions. However, quoting the work of other scholars, the authors highlight the distinction between "divided" and "contested" cities: "the former can arise from antagonism between two or more ethnic or religious groups, while the latter relates to a more fundamental hostility about the ownership and control of the city" (*ibidem:* 55). Other scholars refer to "fractured urban arenas caught up in border and sovereignty disputes as 'frontier' cities" (*ibidem*). The authors provide then an overview of the main understandings and "shades", or variations, of the term "divided city" according to previous literature in the field. Some scholars, for instance, point out the role of wider national conflicts in urban divisions and thus argue that cities such as Jerusalem become "polarized" rather than simply "divided". For instance, Marcuse and van Kempen (2002) mainly use the term "partitioned" to describe socio-spatial divisions in the US. A previous work by Calame and Charlesworth (2011) focuses on cities that have undergone "self-imposed apartheid" due to inter-ethnic civil conflict. Bollens examines various types of divided cities, including those with cultural diversity in multinational societies, those facing intercommunal conflict, those operating as platforms for conflicting sovereignty claims, those within ethnically tense and fragmenting states, and those coping with societal uncertainty and multinational tension. The book thus calls for a reconsideration of planning practices and tools to intervene in such contexts, by means of collaborative and agonist approaches, following the theorization of Jean Hillier.

Scott Bollens is one of the most prominent authors in the field of urban divisions. Along with the work quoted by Gaffikin and Morrissey, two other volumes "City and Soul in Divided Societies" (2012) and "Bordered Cities and Divided Societies. Humanistic Essays of Conflict, Violence, and Healing" (2021) provide a thorough discussion on the role and responsibilities of urban planning and design may help or hinder a healing and recovery process.

"City and Soul in Divided Societies" focuses on nine cities—Jerusalem, Johannesburg, Beirut, Belfast, Nicosia, Sarajevo, Mostar, Bilbao, and Barcelona—addressing "historical, theoretical, and practical issues of urban divisions" as well as providing "a first-person account of conducting interview-based ethnography in these places of ethnic and nationalist polarization" (2012: 5). The viewpoint provided by the book is, thus, very much related to the long experience of the author in these contexts, combining academic research, qualitative interviews and personal observations.

The nine cities are defined as "polarized" rather than "divided", meaning that all these urban contexts "contain a depth of antagonism and opposition beyond what the word 'divided' connotes" (*ibidem*: 5). Polarization occurs at different levels in the urban spaces and political arenas, but also in the everyday living practices of their inhabitants: in studying and engaging with such places, the author states that one can experiences "the extremes of light and darkness" (*ibidem:* 6).

In "Bordered Cities and Divided Societies Humanistic Essays of Conflict, Violence, and Healing" (2021) Bollens returns to the same cities of "City and Soul in Divided Societies". With a similar hybrid language and style, combining ethnographic observations and academic sources, the author adopts Walter Benjamin's guiding approach "that writing about the city should resemble experiencing the city itself, that a discontinuous, fragmented, direct, and immediate writing style more authentically represents the city experience than does a traditional academic structure that emphasizes abstraction and systematizing thought" (Bollens 2021: 3). The book is organized into four main sections, providing a series of short narratives on the nine cities, based on interviews recorded in twenty-five years of research. The narratives are organized around the four themes that structure the volume: Polarized: dichotomization and psychological 'othering'; Emotions: interior emotional world of loss, displacement, and trauma; Place: physical qualities of urban environments shaped by ideological conflict; Time: legacies of past conflict and prospects for the future (ibidem: 2).

By using such an approach, the author identifies the main focus of the book being on "the challenges presented by nationalistic, ethnic, and religious differences and consideration of the other" (*ibidem*: 4), an issue considered as a global concern and that "extends well beyond the nine extremely contested cities that I have studied to encompass numerous cities worldwide that exist in vulnerable and fragile political settings and face major problems of societal unrest and inequalities" (*ibidem*). In this sense, the book recovers the same main issue explored in "City and Soul in Divided Societies", that "divided cities" can be considered as "samples" of urban division that can be found also in other contexts, and thus providing relevant "lessons" for other cities. In addition, Bollens emphasizes the role of the urban scale—even if interlocked with the national level—for his research, focusing on "the spatial location of antagonistic groups, land control and territoriality, development patterns and trajectories, demographic patterns, location and magnitude of urban violence, borders and boundaries, economic relationships, the location and delivery of public services, and the nature of nongovernmental and civil society groups that operate in the city that seek to embed or transcend intergroup differences" (*ibidem*: 5).

One key contribution of the volume regards its capacity to convey not only scholarly meaningful materials and documentation, but also embracing the "the social and urban dimensions of peace as well as its political and institutional aspects" (*ibidem*). Such an approach needs to be grafted into policy-making practices related to peacebuilding, away from a more technocratic approach that proved to be ineffective to tackle urban conflicts and divisions.

The book by Giulia Carabelli "The Divided City and the Grassroots The (Un)making of Ethnic Divisions in Mostar" provides a quite different approach to the ones mentioned so far. The volume is focused on one specific case study, Mostar in Bosnia and Herzegovina, a city that, due to its history, has become "the epitome of ethnic divisions, religious violence and nationalist intolerance" (Carabelli 2018: v). The author's main aim is to engage with the issue of persisting global and local narratives of division that has a long-term horizon in time, space and people's minds. Thus, the book's ambition is "to provide a multifaceted and in-depth understanding of the social, political, and mundane dynamics that keep this city polarized whilst considering the potential that moments of inter-ethnic collaboration hold in reimagining Mostar as other than divided" (*ibidiem*). The attention to everyday cooperative practices to overcome the persisting divisions is one of the main contributions of the volume, which engages with three main fields of study: "ethno-nationally divided cities, urban conflict studies, and the politics of grassroots and movements in the context of socio-cultural segregation" (*ibidem*: vi). Through the case of Mostar, the volume provides critical analysis of urban spaces of resilience and their role in bringing about social change (*ibidem*). However, Mostar represents one of the most quoted case studies when dealing with the issue of "difficult heritage", due to the spatial re-appropriation through the reconstruction of an intentionally destroyed built element of patrimonial value, the Stari Most bridge. When dealing with heritage issues in the case of Mostar, Gustav Wollentz in "Landscape of Difficult Heritage" (2020) outlines the role of temporality in the sense of belonging to the built environment which can overpass the spatial division. Belonging to the pre- and post-war Mostar for local communities can signify a change in the use of the built environment as well as a shift in the narrative of what cultural heritage represents at a specific moment in time. An issue shared by other "divided cities", further tackled in the "Urban Heritage in Divided Cities: Contested Pasts" edited by Ristic and Frank (2019). The book collects a series of essays by different scholars exploring the relevance of contentious urban heritage—defined by the editors as both tangible and intangible elements of the history of a city—in divided urban contexts (Ristic and Frank 2019: 2). In particular, the book's contributions give evidence to "various transformations of urban heritage—appropriating, rejecting, selecting, destroying, preserving, remaking, sharing—and analyze their spatial, social and political causes and consequences for the division and reunification of cities during both wartime and peacetime conflicts" (*ibidem*). In addition, the volume explores ways in which urban identities and places shape each other in contested spaces. The contributions focus on cities from very diverse countries, ranging from the most renowned cases such as Palestine/Israel, Cyprus, and Lebanon, to more un-explored contexts such as India and Colombia. The book delves into the multifaceted role of the urban heritage within divided cities, not only as a focus and a potential object of conflict but also as a socio-political tool for opposition in all analyzed cases. In these contexts, division can be paired with a well-defined physical partition, or with a more unstable and fluid boundary, shifting "through the cityscape" and materializing "through different architectural and urban elements as well as through mental barriers" (*ibidem*: 7).

The main contribution of this volume to the debates on urban divisions regards the re-conceptualization of the notion of urban heritage by shifting the focus away from the traditional understanding of it as a static and unchanging legacy of the past.

Instead, the book seeks to employ the concept of urban heritage as a dynamic and transformable entity that assumes an active role in the critical inquiry, subversion, and transformation of the present (*ibidem*: 10). The essays of this volume illustrate the potential of urban heritage to either divide or bring people together. The chapters serve as a reminder that the agency of heritage is implicated in the ever-evolving power relations in societies, which are sometimes more significantly influenced by the factors and actors on the national/state level than by those on the urban level (*ibidem*).

This very limited selection of volumes corroborates academic positions and criticisms that have started to question the very nature of the term "divided city" and the "taxonomist" approach that has been characterizing the scientific production on this matter in the past decades. Several scholars (Allegra et al. 2012; Carabelli et al., 2021) have, in fact, pointed out the risks associated with the context-free use of the wording "divided city" as well as the recurrence of the same group of cities alternatively re-labeled. Rather, the importance of considering the complexities, peculiarities and differences between cities strongly affected by phenomena of division, segregation, and polarization has started to emerge in order to avoid the over-simplification often associated with contested urban environments, that do not consider the multiple correlations between the historical, spatial, and socio-political dimensions of urban conflicts. Allegra et al. (2012) have listed and criticized the various denominations and definitions of the same cities that have been alternatively "tagged" according to the lens and disciplinary perspective adopted. As we have already observed, the same group of cities—such as Mostar, Sarajevo, Nicosia, Jerusalem, Beirut and Belfast—have been defined as "dual", "divided", "fragmented", "partitioned" if read through the lens of globalization. From a "border studies" approach, adjectives such as "twin", "adjacent", "binational" and "duplicated" are the most recurring ones, related to the same collection of urban case studies (*ibidem*: 563). The issue of how we look at things and phenomena if we change the disciplinary "glasses" is of course not limited to the field of "divided cities" only and represents a crucial bias to be taken into consideration by any collaborative research attempts. At the same time, the realm of "divided cities" studies appears particularly affected as regards the reduction of a "single factor explanation" and consequently to "ideal types" (*ibidem*), failing to recognize and properly convey the complexity of urban identities and imaginaries in cities that have experienced different kinds of conflict and division often overlapped in multiple stratifications in time and space. Despite their different labeling, one aspect becomes common in defining these cities, especially when dealing with the built environment and its meanings, roles and uses in reference to multiple local identities: the role that cultural heritage can play in both ways—partition and reunification.

Such a critical perspective poses a series of questions of primary importance in the context of this volume: how should we consider Nicosia? Is it a "partitioned", "divided", "segregated" or "fragmented" city? What role does the urban heritage of the city play in shaping its (multiple) identities and how, in turn, such identities affect Nicosia's cityscapes and its urban heritage?

How Nicosia's multiple fragilities could be tackled, from the perspective of inter-communal cooperation, while still dealing with a political and ethnic division?

3 Divided, Polarized or Contested? Situating Nicosia's Partition in Time, Space and Imaginaries

In a paper published in 2018, social anthropologist Yannis Papadakis wrote:

> In 1960, Cyprus became an independent republic, the Republic of Cyprus, an outcome that did not satisfy either the Greek Cypriot majority (80 per cent of the population) which strove for Enosis (Union with Greece) or the Turkish Cypriot minority (18 per cent) that demanded Taksim (Partition).
>
> Both ethnic groups continued to pursue their opposed aims after 1960, resulting in interethnic violence from 1963 to 1967 and the arrival in the island of UN peace-keeping forces in 1964. The line dividing its capital, Nicosia, in 1964 was called "the Green Line", as the current division is still also known. In 1974, a group of Greek Cypriot right-wing extremist, with the support of the Greek junta ruling Greece at the time, launched a coup against the island's president, Archbishop Makarios that resulted in a military offensive by Turkey dividing the island throughout. Past and subsequent population displacements rendered the two sides ethnically homogenous. The Republic of Cyprus joined the EU in 2004, which in fact meant that only the Greek Cypriot-controlled southern side became part of the EU, while the Turkish Cypriot north remained outside pending a solution to the Cyprus Problem. The Green Line dividing Cyprus thus became the EU's easternmost border" (Papadakis 2018: 287).

This short account effectively synthesized the events leading to Cyprus' partition and its most recent history, marking a "ping-pong" of responsibilities between the two communities—including the *new* and *old* forms of colonialism that disrupted the island's political stability, and the "missed opportunities" for reconciliation.

But how did the Green Line actually emerge as a persistent geopolitical and spatial device in Cyprus and Nicosia's physical and mental map? The Green Line spans over 180 km across the island, ranging from a thickness of a few meters to various kilometers crossing rural areas, agricultural terrains, the city of Nicosia and other medium-sized and minor urban settlements on both sides. Since the end of the armed conflict, the Green Line is controlled by the UN Peacekeeping Force in Cyprus (UNFICYP), except for the British Sovereign Base Area of Dhekelia, on the eastern side of the island, and the ghost town of Varosha near Famagusta, now under the control of the Turkish military.[6]

The configuration of the Green Line—as many other arbitrary boundaries drawn on maps—represents first of all the "spatialization" of a top-down decision aiming to cope with "the urgency of a compromise, the necessity of war", transforming "the tremor of a human hand" into "urban minefields navigated daily by residents who are unable to avoid them" (Calame and Charlesworth, 2011: 8).

[6] Source: https://unficyp.unmissions.org/about-buffer-zone#:~:text=The%20buffer%20zone%20%2D%20also%20called,is%20a%20few%20kilometres%20wide (Accessed December 2022).

Calame and Charlesworth (2011) suggest, thus, that the Green Line is a "china-graph frontier" which has been designed "by a wax pencil in the space of an hour, remaining split for decades" (*ibidem; see also Strüver* 2018*: 7).*

The authors acknowledge how the drawing of the Green Line occurred as an external act claiming to de-exhalate political tensions and armed confrontations between Turkish Cypriot and Greek Cypriot communities in the wake of anti-colonial struggles and ethnic conflict in the early 1960s. However, if the actual spatial configuration of the Green Line can be read as the product of an external intervention, the process that was brought to the partition cannot be simplified or referred only to such exogenous forces, but to an entanglement of local and global power interplays. As mentioned by Papadakis (2018) ethnic clashes and political turmoil already started in the early 1960s in the context of anti-colonial struggles and independence from Britain. The presence of powerful rulers before and "guarantors" then (the Ottoman Empire/Turkey, Greece, and Britain) and big international organization such as the United Nations and the European Union, combined with nationalist aspirations and narratives, played a key role in the modern and recent history of the island and in the fueling or de-exhalation of the conflict.[7]

Due to the accession of the Republic of Cyprus (RoC) to the European Union in 2004, the Green Line is today a de facto (political) border of the EU with the Turkish Republic of Northern Cyprus (TRNC), a territory which is not recognized by the international community. However, the relationships across the Cypriot divide and its communities do not follow a clear-cut "us-them" situation. As also explained by Strüver (2018), the RoC claims both Greek and Turkish Cypriot citizens that, thus, can have access to EU-wise identity cards and passports.

A different condition regards instead the so-called international communities that live on both sides of the island: since in the TRNC the refugee protection measures are still under development and the RoC has to implement UNHCR and EU regulations, "refugees on the island can apply for recognition of their status through asylum processes in the south only. (...) for the last two decades, the most frequently used route to access refugee protection has been to cross from Turkey into northern Cyprus, then cross the Green Line, and seek asylum in the south. This has rendered the Green Line an important factor in formal and informal policies of refugee protection and border policing" (Demetriou 2021: 7). Such dynamics have also been key in the way in which the RoC and the TNRC have been attracting different communities, generating new space-imaginary relations. Something occurring also in the most remote and rural areas along and inside the Green Line, away from the heavily patrolled and militarized urban core of Nicosia. As recalled in the contribution by Gaeta and Pasqui, "everybody [has] his or her own Cyprus problem" (Mete Hatay interviewed by Horst 2022: 482).

[7] For more detailed and clear explanation of the "Cyprus problem" we refer to the book by James Ker-Lindsay (2011) which responds to a series of crucial questions on the partition and its historical, (geo)political, social and cultural roots.

3.1 *The Green Line in Nicosia: A Thick (Im)permeable Threshold, a Frontier, a Resistant Space and a Place of Resistance*

As the capital city of both the RoC and TNRC, Nicosia represents a central node in the spatial and political dynamics of the island: here the continuity of the urban and social fabric of the city appears radically disrupted and fragmented due to the Green Line. To understand its role in the everyday life of the city and its built environment, it is necessary to first unfold the process of production and the effects of the partition. To do that we refer to the extensive and thorough ethnographic and cartographic work produced by Anita Baskhi and collected in her book "Topographies of Memories. A New Poetics of Commemoration" (2017). The book unpacks the socio-political conditions that brought to the division of the island and of the city of Nicosia. How can a line drawn on a piece of paper become a real border? In responding to this question, Bakshi explores the historical layering of Nicosia's central marketplace of Ermou Street,[8] once the main and most lived commercial area of the city before the partition hosting "a mixture of Greek, Turkish, and Armenian businesses" (Bakshi 2017: 89) and today mostly included within the Buffer Zone.

The morphological configuration of the street is directly influenced by its former use, namely the Pedieos riverbed that once ran through the city center, later diverted by the Venetians when they built the city walls in the sixteenth-century (*ibidem*). The river's morphology still makes its presence visible when heavy rains hit the city, with consequent floods, especially on the eastern side of the city in the area of Famagusta Gate.

Through an intensive collaborative mapping exercise and a collective historiographical narrative based on the voices and memories of several old inhabitants of the North and South Nicosia, Bakshi discloses histories and memories that have been under-rated or neglected by official nationalistic narratives, animated by both sides in the past five decades. Through her exploration, Baskhi is able to surface detailed information about how this part of the city used to be during the most crucial moments of Cyprus and Nicosia's recent history, ranging from the 1950s till 1974, identifying three main phases and related cityscapes: The Everyday City: 1950s; The Devolving City: 1955–1963; and The Enclaved City: 1963–1974.

In these three different "cities in time", whose "boundaries" are not clearly defined but are marked by the intertwinement between top-down political events and social distress, Ermou Street has played a central role, being first a space of encounter and exchange and a platform for social and cross-cultural interactions, then as a place of struggle and conflict, and finally as an element of partition. As a result, one key argument brought by the book suggests that Ermou Street became a spatial device of division, because and foremost it used to be an "interface" between Nicosia's communities (*ibidem*: 120).

---

[8] For a more detailed presentation of this topic, see the Chapter "Tackling Residuality through Nicosia's Market Heritage" in this volume.

The borderline was ultimately created by removing the thresholds of exchange within the city and by hindering those spaces of proximity and "visual familiarity between individuals beyond their ethnic belonging" (*ibidem*: 128). In this sense, according to Bakshi's research, the partition didn't occur as a sudden event, but it was the final step of a process of fragmentation and dissolution of the commercial and social ties, and the related urban fabric and architectural continuity (*ibidem*: 129). Seen from this perspective, Nicosia and its inhabitants have experienced different stages of segregation, enclaving, partition and fragmentation, throughout its recent history. All these urban conditions defined the city as a highly complex environment that cannot be simply labeled as "divided", even if this term has come to overlap with its own identity as a kind of urban marketing strategy implemented by the two municipalities.[9]

3.2 The Role and Conditions of the Built Environment Along and Within the Buffer Zone

Strüver (2018) provides a relevant discussion on the nature of the Green Line, not as a linear element nor an actual international border, but as a thick space in-between different "lines"—the Green Line itself, the Turkish Forces cease-fire line, and the Greek Cypriot National Guard CFL—that have very different meanings for the two communities. "From a [Greek Cypriot] GC point of view, the border is often called the "Occupation Line," referring to the Turkish invasion and dividing the 'free' from the 'occupied' areas of the island, whereas from a Turkish Cypriot (TC) perspective, it is called the "Peace Line" that guarantees rescue and security and thus refers to the Turkish intervention in 1974" (2018: 7).

This is also reflected in the way the two sides of the Green Line are spatially configured. As explained by Casaglia (2020) and how it is possible to observe on-site, the southern side is marked by temporary elements (i.e., tires, permeable metal nets and barrels) indicating the uneven condition of the "occupation line", while the northern side is marked by more stable walls and elements that indicate the line as a protective element. As the essays of this volume will show more extensively, also the areas in proximity to the Buffer Zone on the two sides reflect such different conceptions and perceptions towards the physical partition in the everyday inhabitation practices by the Turkish or Greek Cypriots or even non-Cypriot nationals, as well as different tangible manifestations of the (once) same built environment.

Due to the conflict and its long-term consequences, the area within the Buffer Zone cannot be accessed, if not only occasionally with the authorization of the UN, therefore "buildings cannot be altered or demolished, and the effects of time can clearly be witnessed on their scarred figures" (Bakshi 2017: 89). Because of this state of "crystallization", the built heritage inside the Buffer Zone has been experiencing

---

[9] See, for instance, the homepage of the Nicosia Municipality website: https://www.nicosia.org.cy/en-GB/home/. Accessed July 2023.

progressive decay and collapse. Again, Bakshi well describes the current condition of the area within the Buffer Zone as "a place that has been 'set aside' from the life of the city—its testimonies have been compromised, and its access to time through place is available only through quick glimpses through physical barriers, or through memory" (Bakshi 2017: 140).

In 2013, the Buffer Zone became subject of concern for Europa Nostra that, with the support of local actors, indicated the area as one of "The 7 Most Endangered Monuments and Sites in Europe", due to the advanced material decay of the built environment. This act further stressed the effect that the "inaction" could have had on the entire area of the walled city, north and south.[10] After only nine years, these "cascade effect" can be observed, especially when the authors of this book were challenged to draw the Buffer Zone's south and north limits, using different sources of reference, from online digital tools and municipal geoportal, to various published materials. Each of them depicts a different spatial configuration of the Buffer Zone, while a few of them partially match the in-situ observations and analysis performed by our team. Due to a state of advanced material decay or once a building collapse even partially, the articulation of Buffer Zone itself changes, becoming a terrain of contestation.[11] The "uncertain state" of the Buffer Zone is reflected in the material conditions of the buildings and in the ambiguous situations created while "drawing the division line" such as properties that are cut in two, residual built and urban spaces, uncontrolled partial demolitions behind the walls, and prohibited accesses.

For this and many other reasons, buildings and spaces along and in the proximity of the Buffer Zone have become less and less attractive for the local inhabitants leading to a progressive abandonment of the most central neighborhoods throughout the years. The reasons for such a process are political, but also and mostly stratified in the dramatic and traumatic past of the local communities and perception of the old town's spaces (Casaglia 2020: 62–65), along with the persistent heavy militarization and feeling of unsafety that Cypriots on both sides reported (*ibidem*). It is only in recent years, thanks to the opening of the first crossing point in 2003 (at the Ledra Palace Hotel), the Ledra/Lokmacı Street in 2008 and bi-communal efforts implemented by the two city administrations leading urban regeneration projects,[12] that the quality of public spaces in the area improved, leaving however space for further intervention. The presence of bottom-up and spontaneous renovation initiatives by local organizations and young entrepreneurs involving certain areas in the proximity of the Buffer Zones started to have visible effects. The ground floors have

[10] Source: https://www.europanostra.org/europa-nostra-eib-institute-mission-visits-buffer-zone-nicosia/

[11] For instance the recent diatribe over the Spitfire Café in the Paphos Gate area.

[12] Under the umbrella of the Nicosia Masterplan (NMP), a bi-communal initiative implemented from the mid 1980s by local and international technical teams to deal with common challenges for the city and planning for a future reconciliation. This is one of the most relevant examples of technical cooperation on urban planning matters across in divided territories, also winning a series of international funding and recognition since then. For more details on the NMP please refer to the contributions by Bricocoli and Chrysochou, Gaeta and Pasqui, and the interview with Agni Petridou and Ali Güralp in this volume.

become crucial in hosting social and cultural uses, together with the economic ones, contributing to the reactivation of public spaces and streets. Spontaneous or coordinated, all of these attempts prove how the existing built environment can become a valid local resource, with multiple effects on the social and cultural re-appropriation of the difficult and contested areas of the city. The new *use in time* of the built environment, that passes slowly from abandonment and even resentment towards re-use, even if not by the "community" that identifies with it, outlines the role that the historical architecture in Nicosia can play in defining daily habits and interactions with the tangible and intangible limits of the Buffer Zone.

4 Walking in Nicosia: Constellations of (Extra)Ordinary Spaces Along and Across the Buffer Zone

Yannis Papadakis suggests five paradoxes to unfold the socio-spatial dynamics occurring along the Buffer Zone and envisioning it as a site of: (1) division and contact, (2) conflict and cooperation, (3) security and anxiety, (4) extreme expressions of nationalism and its contestation, and (5) creativity and oppression (Papadakis 2018: 288).

Despite embracing the paradoxical nature of the Green Line, we hereby suggest overcoming binary narratives that might convey polarized readings of the partition. Instead, we propose three "multidimensional" narratives about the city's spaces around and along the Buffer Zone, through some visual notes taken by the authors in September 2021 and March–April 2023.[13] These narratives have been developed thanks to encounters with relevant actors and stakeholders in the city that also accompanied us in a series of collective walks that served as occasions to stop by, observe details of the urban spaces, analyze the features of the built environment and its transformation in time, and listen to stories that otherwise wouldn't be possible to know. Walking is a well renowned and established methodological tool in any empirical research work in urban settings. However, it is crucial to underline, in this context, the collective dimension of such research operations, enriched not only by the personal observation of a place, but also with the interlocutory and conversational opportunity provided by this experience.

[13] As a general premise, the photos hereby presented are not meant as professional products, but instead as notes taken on-site and used as guiding materials for the elaboration of fieldwork outcomes.

4.1 Playgrounds at the Border

Walking along the Venetian Walls on Northern Nicosia, just over Paphos Gate in the southwest part of the walled city, one can encounter the thinnest, almost non-existent section of the Buffer Zone where it comes up to the city walls, disappearing for about 200 m. Until 2003—with the opening of the new crossings in the walled city—it was the only point in the whole island where Greek Cypriots and Turkish Cypriots could look at each other. Here the role of the Venetian Walls as the symbol of the city's historical past—complete as a whole, and without any physical disrupters embodying the complexity of its multi-ethnic communities—becomes manifest. Moving towards the top of the Roccas/Kaytaz bastion on the northern side of Nicosia, there is Sınır Park, close-by the Arabahmet neighborhood, an Ottoman and Armenian legacy which has been in recent years partially regenerated in the framework of the Nicosia Masterplan.[14] On one side Sınır Park, which literally means "border park", overlooks southern Nicosia, while on the other is directly facing the moat where UN watching towers stand out along with the TRNC's and Turkish flags.

Not far away from the park, a row of newly opened cafeterias and restaurants occupy the ground floor of the recently refurbished historic houses along Zahra Street overlooking the renowned Ledra Palace Hotel, on the other side of the Green Line. Down within the moat, a new football pitch shines with its brand-new grass and white football goalposts.

This area appears today as a remarkable place where to observe the stratification of everyday life practices and spaces on the two sides of the Green Line, where public amenities, playgrounds for children and services for the young inhabitants of the neighborhood co-exist with the typical symbols of border patrol in a radius of few meters. All along their circuits, both on the top of the bastions or inside the moat, Nicosia's Venetian Walls host a series of public amenities: football pitches, public parks, gardens and playgrounds, with different conditions of maintenance or conservation of the historical heritage, testifying, on one hand, their recognition as an important and living space of the city on both sides, but also the lack of other green public spaces within the walled city. One of these spaces, as mentioned, is the area devoted to sports and football, as one of the national and city's "pride" for all local communities. Close-by Famagusta Gate on the south side of the Caraffa/ Altun Burcu bastion inside the moat, the Orphea's football pitch has been home for the homonymous football club since 1948. In the area, along Athinas Street, a series of local coffee shops and tavernas provide crucial social sites for local inhabitants, especially older football fans, while on the other side of the football

[14] As mentioned, the NMP allowed for the cooperation among the town planning departments of North and South Nicosia on joint projects, giving priority to the old town and the rehabilitation of neglected historical neighborhoods inside the city's wall. The first implementation phase of the Masterplan started in 1986. This included twin projects for the rehabilitation of two areas located along the buffer zone: Arabahmet in the northern part of the city and Chrysaliniotissa in the south (Petridou 1998).

Fig. 2 The Caraffa/Altun Burcu bastion at Paphos Gate. Here the Buffer Zone is almost non-existent and coincides with the walls' surface. Photo by Alice Buoli, 2023

pitch, Palouriotissas Lyceum school complex and related sports facilities directly face the Buffer Zone (Figs. 2, 3, 4 and 5).

4.2 *Transnational Social Spaces*

Along with internal refugees' displacement and the enclaving season that occurred as a consequence of the 1960s conflict, Cyprus and Nicosia have become transit or arrival places for many international migrations flows since the 1990s (Demetriou 2021). The impact of the migratory phenomenon over the city has been primarily related to the housing demand—that induced the occupation of some of the empty buildings of the old town (Casaglia 2020: 60; 64)—as well as to the emergence of a network of temporary or more stable spaces and services devoted to asylum seekers and other populations.

These range from support facilities for refugees and migrants—managed by international and local NGOs, religious orders or by the UN—to a series of "transnational

Fig. 3 Children playground at the Sınır Park in North Nicosia. Photo by Alice Buoli, 2021

social spaces" (Faist 2000)[15] in the city (i.e., shops, temporary markets, spaces for cult, or internet cafes) that provide a network of services and products for a variety of inhabitants.

On the southern side of Paphos Gate nearby the Catholic Church of the Holy Cross and the Vatican Embassy in Cyprus, a tiny temporary market is organized on Sunday morning by Southern-East Asian women selling typical products from their home countries, benefiting from the coincidence with the morning mass.

This used to be a very important area for religious groups before the partition: along the same street are located the Holy Cross church and the Maronite church in the South, up north in the Arabahmet neighborhood there are the Armenian church and monastery, as well as the Arabahmet mosque.

On the other side of Pafou Street, the walls of the former UN 65 headquarter building (now abandoned) a series of posters advertise services and meetings for prospect entrepreneurs and for migrant women. Just few meters away south a bill-board advertises the presence of a religious order, the Sri Lankan Sisters of Our Lady of Perpetual Help Nicosia.

[15] Thomas Faist defines "transnational social spaces" as the "sustained ties of persons, networks and organizations across the borders of multiple nation-states, ranging from weakly to strongly institutionalized forms" (Faist 2000, p. 2).

Fig. 4 The new football pitch in the Buffer Zone in North Nicosia seen from Zahra Street. Photo by Alice Buoli, 2023

It is no coincidence, therefore, that Caritas Cyprus—engaged in humanitarian activities and support for migrant populations—decided to settle in this area, due to the density of religious 'landmarks' and in proximity to the city's main bus station.

Spontaneous religious practices are another relevant (visual and acoustical) presence in the same area, where a series of ground-floor spaces and storefronts are used for cultural and devotional activities by various groups. Their presence is particularly evident on Sunday morning when music and chants fill in the streets around Arsinois Street on the western side of the city. As noted in the contribution by Giambruno and Oteri, most of religious heritage in Cyprus was affected by the partition, following the internal displacement of Greek Cypriot and Turkish Cypriot communities. On one hand, the south-north divide triggered the abandonment, transformation, and re-use of various cult buildings. On the other hand, the influx of new communities in recent decades, resulting from different "displacement" dynamics such as transnational migratory phenomena, led to the spontaneous development of novel religious spaces and practices within the built environment.

These spaces can be considered as nodes of more extended networks of solidarity and self-support that migrant communities have been implementing in the city, especially in the old town on both sides. For this reason, many human rights associations or migrants-related services have found homes in neglected spaces in proximity to the Buffer Zone.

Fig. 5 The fence and billboards of Orphea football club and in South Nicosia. Photo by Alice Buoli, 2023

One interesting case is the Nicosia Dignity Center, a project implemented by the international NGO Refugee Support. The local team has completely renovated and refurbished a 3-story building in the city center (in Perikleous Street) and provides first-aid services to refugees, most of whom come from the Pournara Emergency Reception Center, located a few kilometers away from the capital. On the north side of the city, the Refugee Rights Association[16]—the only organization working with refugee protection in the TRNC—has its offices nearby the Ledra Palace Hotel checkpoint.

At the same time, it must be mentioned that—due to the property controversies that are still persisting as a consequence of the conflict and the partition—many buildings in the old town cannot be fully renovated or equipped with adequate services and basic sanitary and energy infrastructures. This—combined with emerging hints of "gentrification"[17] whose extension and impacts are yet to be assessed—is making the

[16] https://mhdkibris.com/?lang=en. Accessed April 2023.

[17] Here the term "gentrification" is intended in its general meaning as a process of social and demographic change within a certain neighborhood or urban area as a consequence of urban or architectural transformation (either as public policies and planning or private initiatives), usually followed by a rise in the value of building properties and housing rents. In Nicosia, the stratified social composition of the central neighborhoods of the city, also as a consequence of the partition, the term "gentrification" might be not fully adequate to describe the ongoing process of

Fig. 6 Entrance to the Dignity Center Nicosia. The picture was taken on Sunday morning when the offices are closed to the public. Photo by Alice Buoli, 2023

material and social conditions of Nicosia's central neighborhoods quite composite, with the presence of the most disadvantaged populations in the less desirable, and likely the most decayed, housing units and buildings (Figs. 6 and 7).

4.3 Wall(paper)s

Walking along and in the proximity of the Green Line on both sides, and in some other spots[18] within the city center(s), it is impossible not to notice the presence of a rich production of artworks on the city's walls.

---

demographic transition in light of recent regeneration processes. For more detailed discussion and through literature review on this term please see Gaeta (2008).

[18] In North Nicosia the most relevant areas for street art are Iplikpazari, Karamanzade, and Selimiye; while in South Nicosia they are: the so called "Graffiti Garages", the Archbishop, Chrysaliniotissa neighbohood, the Stasinou Mural Hot Spot, Famagusta, Digeni Akrita, Esperidon, Tophane, Spyrou Kyprianou. A map of these areas is available at: https://vagabundler.com/cyprus/streetart-map-nicosia/ (Accessed December 2022).

Fig. 7 The Cyprus Church storefront space in South Nicosia. Photo by Alice Buoli, 2021

Most of these art pieces are charged with political meanings and messages strongly engaged with the "Cyprus problem" as well as with youth cultural movements, and LGBTQI+ and feminist struggles.

Pafsanias Karathanasis interestingly observes the link between the emerge of street art practices and the liminal nature of Nicosia's old town, allowing for the emergence of the spaces and practice of (artistic, social, and political) free expression: "in a liminal space like the Old Town of Nicosia, it is not only social but also political, religious and ethnic boundaries that become blurred, and in becoming blurred, they also come to be contested. In such an area, the relation between the material characteristics of the liminal space, and the political and activist interventions within it, reveals alternative views to official narratives of both the border and the everyday socio-political crisis of Cypriot society" (Karathanasis 2021: 207).

At the same time, street art is by definition a mode of expression occurring in the city and in its public spaces: for this reason, is always context-sensitive reacting to the urban elements and presences where they are displayed, for instance in proximity to border fences or checkpoints. Again Karathanasis (*ibidem*) argues the mutual influence between the political interventions in the public space and the production of political spaces within a certain urban domain, in this case already charged by strong political connotations and meanings (Figs. 8 and 9).

Fig. 8 Feminist graffiti in the area of Faneromeni Square. This artwork is located on the wall of Ta Kala Kathoumena one the first cafes that opened in the walled city in the late 1990s. Photo by Alice Buoli, 2021

5 Conclusive Remarks and Open Discussion: Designing with Territorial Fragilities in Nicosia

Due to its "frozen", contested, and uncertain status, overlapping with the difficulties in accessing its built heritage, the Buffer Zone is still considered one of the most vulnerable and "marginal" areas in Cyprus and in Nicosia. Once a vibrant and productive interface among the city's communities, today the area within and along the Buffer Zone remains an "open wound" in the social, political, and urban fabric of Nicosia. Since the mid 1960s, as a result of anti-colonial struggles, inter-ethnic conflicts, and geopolitical interferences, the partition has, in fact, triggered a process of progressive abandonment and decay of the built environment and social life along the Buffer Zone. Issues that have been amplified, in recent years, by safety threats to the heritage stock (i.e., collapses, uncontrolled demolitions and seismic risks), climate-related crises (i.e., water scarcity and loss of biodiversity) and social distress (i.e., refugees' crisis and "enclaving").

However, within and despite such "fragile", "liminal", and "peripheral" conditions, through time, new social spaces and practices have emerged along the Buffer Zone and in the walled city, gradually surfacing the potential of this area as a renewed "interface" across the divide. Similar forms of re-appropriation and re-use of the

Fig. 9 Street art in the Bandabulya market area. Photo by Alice Buoli, 2021

existing built environment, vacant lands, and public spaces can be observed on both sides of the city—certainly with different paces and funding sources and volumes—thanks to the initiative of creative communities, youth and female entrepreneurship, human rights, LGBTQI+ activism, and migrant-led organizations. Members of local and international communities have become in recent years the safeguarding actors of the built environment through the simple act of re-dwelling and re-using residual spaces and abandoned buildings.

It is not by chance that the most important bi-communal planning initiative, the Nicosia Masterplan, identified in its early stages in the 1980s the Buffer Zone and the walled city as the main urban regeneration priorities to be tackled in view of the future reunification of the city (or the persistence of the status quo). After more than forty years after the NMP was kicked-off, a common sewerage system, shared water and waste infrastructures, housing, public spaces, cultural heritage and basic collective services are still today the main concerns and key strategic axes on which urban planners, architects, and local administrators confront on a daily basis on both sides.

The essays and projects developed by our colleagues and presented in this volume give evidence to the relevance of everyday life practices and interactions as the terrain on which understanding and imaging "communication bridges" between the communities inhabiting the walled city.

Designing "with" territorial fragilities in Nicosia entails disclosing and reconsidering the role of "ordinary" infrastructures and spaces (i.e., playgrounds and gardens, markets, public fountains, cultural spaces, etc..), as well as common heritage sites, as support on which co-presence, visual recognition and proximity between individuals can occur, despite and beyond the partition.

References

Allegra M, Casaglia A, Rokem J (2012) The political geographies of urban polarization: A critical review of research on divided cities. Geography Compass 6(9):560-574

Bakshi A (2017) Topographies of memories. A new poetics of commemoration. Palgrave Macmillan, Cham

Bollens SA (2012) City and Soul in Divided Societies. Routledge, Abingdon

Bollens SA (2021) Bordered Cities and Divided Societies: Humanistic Essays of Conflict, Violence, and Healing. Routledge, Abingdon

Calame J, Charlesworth E (2011) Divided Cities: Belfast, Beirut, Jerusalem, Mostar, and Nicosia. University of Pennsylvania Press, Philadelphia

Carabelli G, Djurasovic A, Summa R (2021) Challenging the representation of ethnically divided cities: perspectives from Mostar. In: Carabelli G, Djurasovic A, Summa R (eds) Challenging the Representation of Ethnically Divided Cities. Routledge, Abingdon, pp 1-9

Carabelli G (2018) The Divided City and the Grassroots. The (Un)making of Ethnic Divisions in Mostar. Springer, Cham

Casaglia A (2020) Nicosia beyond partition: Complex geographies of the divided city. Unicopli, Milano

Demetriou O (2021) Migration, Minorities, and Refugeehood in Cyprus: A View of the Citizenship Regime through the Lens of Displacement. In: Jahrbuch für Europäische Ethnologie. Brill / Schöningh, Leiden, pp 91–120

Faist T (2000) Transnationalization in international migration: implications for the study of citizenship and culture. Ethnic and Racial Studies 23: 189–222. https://doi.org/10.1080/014198700329024

Gaeta L (2008) L'incerto stato della gentrification. Un sommario della letteratura. Bollettino della società geografica italiana 1(2): 449–462

Gaffikin F, Morrissey M (2011) Planning in divided cities. John Wiley & Sons, Hoboken

Gilloch G (2013) Walter Benjamin: Critical Constellations. John Wiley & Sons, Hoboken

Interviewed by Cindy Horst (2022) Creating a Third Space in the Cyprus Conflict: Mete Hatay. In: Tønnesson S (ed) Lives in Peace Research. The Oslo Stories. Evidence-Based Approaches to Peace and Conflict Studies, vol 3. Springer Nature, Singapore.

Karathanasis P (2021) Writing in a City in Crisis Stencil Graffiti in the Old Town of Nicosia. In: Campos R, Pavoni A, Zaimakis Y (eds) Political Graffiti in Critical Times The Aesthetics of Street Politics. Berghahn, New York / Oxford, pp 201–224

Ker-Lindsay J (2019) The Cyprus Problem in an era of uncertainty: establishing a culture of engagement. PRIO Cyprus Center Report 5. PRIO, Cyprus

Marcuse P, Van Kempen R (eds) (2002) Of States and Cities: The Partitioning of Urban Space. Oxford University Press, Oxford

Papadakis Y (2018) Borders, paradox and power. Ethnic and Racial Studies 41(2), p 285-302. https://doi.org/10.1080/01419870.2017.1344720

Petridou A (1998) Nicosia: Perspectives for Urban Rehabilitation. Journal of Mediterranean Studies 8(2):350-364

Ristic M, Frank SF (2019) Urban Heritage in Divided Cities. Routledge, Abingdon

Silver H (2010) Divided cities in the Middle East. City & Community, 9(4):345-357. https://doi.org/10.1111/j.1540-6040.2010.01348.x

Strüver A (2018) Europeanization in Cypriot Borderscapes: Experiencing the Green Line in Everyday Life, Geopolitics 25(3):609-632. https://doi.org/10.1080/14650045.2018.1550390
Wollentz G (2020) Landscapes of Difficult Heritage, Palgrave Studies in Cultural Heritage and Conflict, Palgrave Macmillan.

Shared "Values" in Divided Contexts. Some Reflections on the Role of Urban Heritage in Cyprus

Maria Cristina Giambruno and Annunziata Maria Oteri

Abstract The essay proposes a focus on the role of urban heritage in divided settings. The purpose of the article is to analyze two different perspectives through a preliminary critical overview of the law and initiatives defined at international level to protect the cultural heritage in conflicted areas since WWII. The urban and architectural heritage as a peacemaker or, on the contrary, how urban, and architectural heritage can become the symbol of conflicts and divisions. The current war in Ukraine shows again how symbolic buildings and urban ensembles can become the target of vengeful actions, which seriously affect the identity and sense of belonging of the communities who suffered wars and conflicts. The essay will focus on some specific case studies, analyzing how buildings and sites once considered symbolic and relevant in terms of history, memory, and identities can become "difficult heritage"; the representation of divisions, defeats, and subjugations. In the last part of the essay, a specific reflection on the case of Nicosia in Cyprus will be analyzed, considering the current perception of buildings and sites, once shared values of multicultural communities now divided by a physical, political, but also a cultural border.

1 The Architectural Heritage in Conflict Areas. From Disallowed to Difficult Heritage

Memory, heritage and identity are concepts that are intrinsically bound to the meaning of cultural heritage as "a group of resources inherited from the past which people identify, independently of ownership, as a reflection and expression of their constantly evolving values, beliefs, knowledge and traditions" (Council of Europe 2005).

Thus, tangible and intangible remains represent the memory of a people, the heritage that past generations bequeathed to the next ones, the representation of a

---

M. C. Giambruno (✉) · A. M. Oteri
Department of Architecture and Urban Studies, Politecnico Di Milano, Milan, Italy
e-mail: mariacristina.giambruno@polimi.it

A. M. Oteri
e-mail: annunziatamaria.oteri@polimi.it

© The Author(s), under exclusive license to Springer Nature Switzerland AG 2023

A. Buoli and O. C. Ţiganea (eds.), *Territorial Fragilities in Cyprus*, Research for Development, https://doi.org/10.1007/978-3-031-36076-3_5

community's identity. As they are considered "properties" of cultural heritage, it is evident how they can change during a war period as well as in conflicts due to religious or cultural reasons, facing evident criticalities in terms of acknowledgement, preservation, and conservation. Furthermore, difficulties appear if also considered the role they play as messengers of sometimes even agonizing memories and legacies, or to establish, claim or, quite the opposite, destroy local identities through the manipulation of its meaning.

The past architecture, the milestone of memory, as a tangible sign of facts and events, often considered as a "done on purpose monument" of a historical period and civilization, perfectly represents the possible impact of a conflict or a crisis on cultural heritage. This may happen when the heritage embodies such a high symbolic value for the enemy, whose main purpose is to destroy it, to erase its memory, to weaken people's resistance or, quite the opposite, when no heritage value is officially acknowledged.

In the first case, we can still remember the destruction of the Stari Most in Mostar in 1993, an event that had a huge impact on the spirit and memory of the Muslim part of the Bosnia Herzegovina's population. The bridge was intentionally destroyed with the main effect of canceling a tangible piece of local cultural identity and distressing the population. Another well-known case related to contemporary armed conflicts is that of the Buddhas of Bamiyan, in Afghanistan, destroyed by the Taliban in 2001. In this case, the Buddhist population, almost inexistent in the country at that time, was not in the Taliban's sights, but their attack was a way to show their power to the international community that tried to preserve and defend the antique sculptures and, subsequently, to the Afghan population, as warning sign to prevent any future attempt to rebel against the new regime.

At the same time, during the current and dramatic war in Ukraine, cultural heritage is intentionally destroyed every day, as the Ukrainian government states. Even if there is no objective evidence—as the war is still ongoing, restricting the access and the accomplishment of a well-defined territorial survey—the destructions target not only the Nation as an institutional organism, but moreover, the national identity of the people. In June 2022 the number of damaged or destroyed cultural sites amounted to 152 (OSCE 2022), in January 2023, according to the website "Destroyed cultural heritage of Ukraine" by the Ministry of Culture, 553 buildings and objects of cultural heritage or belonging to cultural institutions have been bombed.[1]

If historical buildings are not officially considered part of the cultural heritage, due also to ideological reasons, they may be impertinently used for military and strategic purposes. As a result, this underestimated architectural legacy becomes a depository to store weapons and ammunitions, a place to install heavy artillery or missile systems because of its strategic position to attack the enemy. This is the case of the Kosovo Museum in Pristina, used as anti-aircraft artillery stationing, the prehistorical site of Azykh Cave in Azerbaijan, transformed into a depository to store ammunitions, the Malwiya minaret in Samarra, whose upper part was detonated as it

[1] Destroyed cultural heritage of Ukraine. https://culturecrimes.mkip.gov.ua/. Accessed April 2023.

could be used as possible anti-sniping stationing by the US army (Maniscalco 2006a, b: 24).

Another case of destruction due to strategic reasons is the oldest part of Nablus, in Palestine, a wide and stratified historical center, composed of several small buildings and big noble palaces built using clear stone that overlooks narrow streets, where, according to the Israeli government, terrorists, i.e., the members of the Palestinian Resistance, hid. Thus, the old city of Nablus, where, for centuries, Arabic and Samaritan Hebrew population lived peacefully together, was damaged both during the second Intifada, when tanks crossed the city destroying its historical tissue to expand and make this labyrinth of streets more controllable, and during the third Intifada, when both historical soap factories, symbolizing an old production site of the city as well as the main earning source for a population experiencing a dramatic war and its consequent economic crisis, were destroyed. As a result, the architectural heritage of the city has been extremely damaged during the different conflict phases and not only because it was directly bombed.

Because of an inadequate process of emergency structural interventions accomplished by a low-qualified staff that just has a few tools available, architectural heritage may be damaged by the same protection interventions whose main aim is to preserve them. Furthermore, because of its reduced load bearing capacity during bombings, architectural heritage becomes more fragile when tanks cross the streets, as well as because of wind and earthquakes, even if of low magnitude (Maniscalco 2006a, b: 15).

As the architectural heritage is often abandoned after bombings or after the most violent crisis moments of the armed conflicts, and the population moves to other parts of the city—as in the case of the Buffer Zone of Nicosia—historical buildings, whose roof tops, doors, and windows have been destroyed, are even more exposed to atmospheric agents with a consequent acceleration of degradation phenomena due to weathering. Indeed, the end of the most violent phases of a conflict doesn't necessarily mean that cultural heritage does not run any more risk.

Abandonment and vandalism just represent two of the possible risks that cultural heritage runs in the period immediately after the crisis, as there is another risk whose impact is significantly longer in time, and it is intrinsically bound to memory, identity, and material legacy of the aforesaid population.

Indeed, cultural heritage, more specifically the built one, can be considered the messenger of a significant inheritance, characterized by a controversial and divisive memory for civil society. This is the case of all those architecture and urban legacies dating back to totalitarian regimes of the twentieth century in Europe. It deals with a "difficult heritage", as it is specifically defined by Macdonald in 2009. At the beginning, this definition associated the architecture dating back to the Nazi years; afterwards it was used to refer to all built remains of various totalitarian regimes, also involving, during the 1990s, the legacy of the Socialist Realism.

"Difficult heritage" is part of the built legacy that is "recognized as meaningful in the present but that is also contested and awkward for public reconciliation with a positive, self-affirming contemporary identity. 'Difficult heritage' may also be troublesome because it threatens to break through into the present in disruptive ways, opening-up social divisions, perhaps by playing into imagined, even nightmarish, futures. (…) Doing so highlights and unsettles cultural assumptions about and entanglements between identity and memory, and past, present, and future. It also raises questions about practices of selection, preservation, cultural comparison and witnessing—practices which are at least partly shared by anthropologists and other researchers of culture and social life" (Macdonald 2009: 1).

Furthermore, it was defined as "dissonant heritage" a divisive element that should be recognized as heritage itself, whose perception changes in different cultural groups, creating, as a result, a conflict with all values of contemporary society (Tunbridge and Ashworth 1996). It is a definition referring to architectural works dating back to totalitarian regimes, becoming part of the architectural preservation practice and theory over time (Carter and Martin 2019), that, anyway, can trigger, after almost one century, dramatic memories and divisive debates, as in the case of the question coming from the New Yorker in 2017, when Ruth Ben-Ghiat asked: "Why are so many fascist monuments still standing in Italy?".

Thus, this definition may be helpful to understand the role and meaning of cultural heritage in those territories characterized by a physical, religious, social, and cultural division, as in the case of Cyprus and, more specifically, the city of Nicosia, the last European capital city crossed by a tangible border, as it is well known.

2 From Conflict to Reconciliation. The Peace-Making Role of Cultural Heritage in the International Debate

In 2015 UNESCO adopted a strategy[2] to strengthen the preservation of cultural heritage and promote cultural pluralism in case of armed conflicts. "To build peace in the mind of men and women" is the main purpose of the elaborated document, to make people more aware of its importance and increase governments' capacity to prevent and reduce any loss of cultural heritage during wars and conflicts. Another more ambitious goal of this strategy is that preservation of cultural heritage becomes part of humanitarian actions and of a not less important peace-making process (UNESCO 2015). As well known, it does not deal with the first UNESCO's initiative whose purpose is the preservation of cultural heritage threatened by war and conflict destructions. This interest is already evident in the first measures that were adopted at the international level starting from the end of the nineteenth century and, more specifically, the 1954 UNESCO Convention, known as the Hague Convention, as well as in the legislative framework of the following period. Even if all

[2] *Reinforcement of UNESCO's action for the protection of culture and the promotion of cultural pluralism in the event of armed conflict*, 2015, https://unesdoc.unesco.org/ark:/48223/pf0000235186. Accessed February 2023.

passages are not described, as they have been already widely analyzed[3] (O'Keefe 2006), the current goal is not to pay specific attention to all those measures that were adopted to preserve cultural heritage in case of conflicts, i.e., all measures to preserve these goods at physical level and that didn't work properly in many cases. On the contrary, with reference to the case of Cyprus, the main purpose is to try to show how the attention initially paid to physical preservation of that part of cultural heritage having a worldwide value was then paid to a more conceptual dimension, bound to the universal value that this heritage has for the international community to define cultural, social and religious identities of each nation, but, more generally, in terms of cultural enrichment. Therefore, from this point of view, our past heritage, apart from where, how, and when it was established, represents a key tool to prevent any deviation towards different forms of incivility that usually cause conflicts.

The role that cultural heritage played as a peacemaker seems to find a definition over time, at least indirectly, not only at international level, but also, and first, in the legislative framework dealing with practical cases of voluntary destruction (i.e., when there were no military reasons) of buildings having a specific representative value in the history of the nations that were involved in these conflicts (Baroncini 2019). This was due to the destruction of significant parts of cultural heritage (e.g., in Mali, Syria, Iraq), because of different attempts of abuse of power against those populations experiencing conflicts.

The 1954 Hague Convention establishes for the first time that any damage to cultural heritage, irrespective of the people it belongs to, is a damage to the cultural heritage of all humanity, as culture has no borders. More specifically, this is the case of cultural heritage, part of dedicated lists, that have "enhanced protection" as it is considered of "the greatest importance for humanity" (Second Protocol to the Hague Convention 1999). It is known that these principles, triggering anyway debate on this theme, were often considered ineffective practical tools, as already explained in the first part of this essay. A more effective tool, at least to define all aspects connected to this problem, is represented by a specific legislative framework resulting from the processes against people that destroyed cultural heritage. For instance, the International Criminal Tribunal for the former Yugoslavia, established after the bloody conflict dating back to 1991, considers the destruction of cultural heritage a war crime.

The idea to preserve the "right to culture", that implicitly involves the "right to have a historical identity of a people [...] and the interest of all humanity to preserve cultural heritage beyond political differences" emerges from the process documents (Roversi Monaco 2019: 36). Among all cases analyzed by this Tribunal—also considering the verdicts that have been the object of much debate on both sides—the destruction of the city center of Dubrovnik, at that time already listed as a protected UNESCO site, is significantly relevant. This violent action was convicted as

[3] There is a significantly wide international bibliography, especially at juridical level, as well as numerous online portals that are now available, collecting and discussing about the most important initiatives on this subject. A useful reference source is the Unesdoc Digital Library by UNESCO (http://unesdoc.unesco.org/).

its consequence is an irreparable and voluntary loss of value of the historical city that no one can recover through regenerative restoration works, as the buildings that were rebuilt after their destruction show. Despite a minimum penalty imposed on Miodrag Jokić, one of the main accused people, in comparison with the serious destruction, the Tribunal recognized that the damage didn't just refer to the physical destruction of the historical site, but also to the loss of important intangible values connected to religion, education, arts and sciences. "The crime was particularly serious in the present case—as we read in the sentence—because the Old Town of Dubrovnik was, in its entirety, listed as a protected UNESCO site. Residential buildings within the city were therefore specially protected, together with the rest of the site, as an outstanding architectural site illustrating a significant stage in the history of humankind".[4]

The destruction of the Sarajevo's Library as well as the Mostar's Bridge had similar effects, as already said. In the first case, even if the crime was not effectively convicted, the Tribunal recognized the intention to destroy Bosnia's cultural roots through the destruction of a consistent book legacy of patrimonial values. In the case of Mostar, the Tribunal established, with reference to the violation of the proportionality principle, that the consequences for the Muslim community were out of proportion if compared with the advantages deriving from its destruction, pushing towards complete isolation of a one defined community.

Religion was the reason for the more recent destruction of the Timbuktu Mausoleum by Muslim integralists, whose leader was Ahmad Al Faqi Al Madhi; they were generally considered pilgrimage and devotion sites for Malian population against any kind of more integralist doctrine. In this case, the verdict is emblematic, as the International Criminal Court recognizes on this occasion the severe damage to Malian population in terms of violation of their identity and religious freedom and it introduces, for the first time, the concept of crime against cultural heritage. Even if it's not possible to compare it with crimes against persons, it represents de facto a significant step forward in terms of acknowledgement of cultural heritage by international courts of justice as an element to be included among possible "victims" of this crime. This is a significantly more important fact than the generic reference to vandalic actions, as the aforesaid 1999 Protocol indicates. Furthermore, a very interesting aspect in this context is the implicit acknowledgement of the strong bond between a community and its cultural heritage. Indeed, the destruction of this Mausoleum affects the religious dimension of the local community—that had no possibility to profess freely its faith—as well as the cultural one that refers to the educational value of cultural heritage, but also the social dimension, as maintenance of these sites was carried out by the same community (Carstens and Varner 2020). Thus, as the verdict concludes, as they were included in the UNESCO World List, the loss of these architectural works had a huge impact on the community: "The mausoleums of saints and mosques of Timbuktu were an integral part of the religious life of its inhabitants and constitute a common heritage for the community. They reflected their commitment to Islam and played a psychological role to the extent of being perceived as protecting the people of Timbuktu. Furthermore, all the sites but one, were UNESCO World

[4] https://www.icty.org/en/sid/8448. Accessed February 2023.

Heritage sites. As such, the attacks on them appear to be of particular gravity as their destruction does not only affect the direct victims of the crimes but also people throughout Mali and the international community".[5]

Therefore, as we can conclude by reading official verdicts and filed defensive memorandums—all available online—the intention is to recognize the strong bond between a community and its cultural heritage; as a result, the destruction of cultural heritage, even if there are no victims among the involved population, represents anyway a severe violation of humanitarian rights of the directly involved people as well as of the community as a whole.

However, even if there is no explicit reference to the role played by cultural heritage as peacemaker, we can say, despite some forcing, that the acknowledgement of the value of cultural heritage entails the implicit admission of its importance as a source of nourishment for people, an effective tool against cultural and social deterioration. Thus, it's a powerful tool against any form of violence, including war. Anyway, more recent provisions let the conciliatory role that cultural heritage may play, as well as its importance as civilization driving force, emerge more widely. In 2017 the United Nations adopted a Declaration (2347) that was immediately considered an unprecedented victory, whose main purpose was to blame the severe terrorist attacks by Muslim fundamentalists against non-Muslim religious heritage. It deals with an exhortation to all governments to include conservation of cultural heritage among priority safety actions, as it symbolizes unity, a centuries-old dialogue among different cultures, that, as a result, may be an easy target to trigger conflicts. Based on such an immense cultural heritage all over the world, destroyed in these last years, even if there were no military reasons, the idea that cultural heritage is "politically sensitive" takes shape (Council of Europe 2021a, b: 4) and, as a result, it may be ideologically used "against" someone by destroying it. Nevertheless, as it clearly emerges from some reflections in the European Union, it may become a way to heal wounds caused by conflicts and divisions, spread peace and promote peoples' growth from a cultural, civil, and social point of view. This may be considered the result of a stratification process involving different civilizations that are almost always interdependent and, thus, if it's considered part of a shared memory heritage, whose global acknowledgement is necessary to build a common future. Indeed, for this reason, international organizations for conservation, the European Council as well as governments, in general, insist on sensitization of young people to "heritage education" (Council of Europe 2021a, b) and on the necessity to adopt a people-centered approach to preserve cultural heritage (ICOMOS 2015).

[5] https://www.icc-cpi.int/news/icc-trial-chamber-viii-declares-mr-al-mahdi-guilty-war-crime-attacking-historic-and-religious. Accessed February 2023.

3 Architectural Heritage in Divided Cyprus

As we can clearly see in the capital city Nicosia, the architectural heritage of Cyprus is the result of a stratification process involving populations and cultures that crossed the island over the centuries. Remains dating back to the Byzantine and Venetian periods, then to the Ottoman dominion and the protectorate of the British Empire, are visible in both parts of the now divided island. It deals with a stratified architecture in both parts of the territory that indiscriminately represented the history of a community, at least till the 20s of the Twentieth Century, when, while political pressure towards Greece rose, public and scholastic buildings inspired to a Hellenic neoclassical "style" were built (Balderstone 2009).

After the Turkish invasion, the condition as well as the perception of historical architecture changed. At that time, because of such a clear separation of the two main communities, the shared past heritage represented just one part of Cyprus's population, suddenly becoming a "difficult heritage" for all people living in each of the two territories of the island. Of course, this is primarily the case of religious heritage (Fig.1). Churches, monasteries and mosques were the first buildings that immediately experienced the split, because of the forced population transfer from one side to the other of the border, because of new inhabitants coming from Anatolia on the north, and the consequent occupation of abandoned villages and houses. In some cases, in the northern part of Cyprus many Christian religious buildings, churches, chapels and monasteries were vandalized or used for other purposes. Some of them became mosques as well as warehouses, shelters for cattle, resorts, depositories and military complexes, as it happened with the Panagia Acheiropoietos Monastery that only recently was given back and used again for civil purposes and now needs urgent and accurate conservation interventions (Verdiani and Camiz 2020).

In the southern part of the country, some mosques experienced the same situation, being abandoned or occupied as depositories for agricultural purposes.

Of course, the years immediately after the most intense phase of the conflict represented the most critical moment for religious heritage, as well as more in general for cultural heritage. As villages were occupied by refugees from both parts, toponymy changed, and commemorative monuments were replaced—the first elements to be modified when political regimes change—cemeteries were abandoned, and buildings were transformed. However, the difficult role that cultural heritage and, more specifically religious heritage plays in case of political or cultural conflict goes beyond the most intense crisis phases and becomes a long-term element.

Many years after the physical division of the island, the establishment of checkpoints in Cyprus in 2003—an evident first small step to normalize the possibility to move—made the meaning of "difficult heritage" of the so-called monuments important again. As, after years, both communities could see their villages and visit their religious buildings, that, as we said before, were abandoned or used for new and often not appropriate purposes, a new intercultural conflict arose with the direct involvement of the political actors that influenced the public opinion, nationally and internationally, as in the case of Greek-Cypriot religious buildings from Northern part of the island. On the other hand, as churches became mosques, some small devotional actions, i.e., to light a candle or put a sacred image at the entrance, pushed the Muslim community not to go there anymore and, as a result, the Turkish government built new and monumental mosques next to them. It was a vicious cycle, where all those churches that were no more used by the Muslim community to pray experienced degradation and abandonment once again; furthermore, the Greek community perceived the construction of new mosques as another Turkish invasion (Constantinou Costas et al. 2012).

This case emblematically outlines the thin border between two possible roles that architectural heritage could play in conflict areas. On the one hand, it is a pacifying force that triggers virtuous reconciliation processes, on the other, it is a divisive element that strengthens this idea of division. However, it must be said that, over the years, both in the northern and the southern part of the island, churches and mosques were preserved, even if it probably dealt with a measure adopted by both parts to spread tolerance and to open to the "other" community of the island.

On the contrary, without analyzing the quality and effectiveness of such conservation interventions on architectural heritage, in the capital city, they seem to show a still timid but encouraging sign of how past remains may become an element of cohesion. In both parts of the city, thanks to a bi-communal initiative—Nicosia Masterplan—twin interventions have been promoted: for instance, the pedestrianization of the central street crossing and cutting across the physical border between the "two Nicosia" (Ledra Street), or the recovery of buildings dating back to the Ottoman and British period positioned on the "opposite" or "other" side of the Buffer Zone. Indeed, started in the 1980s, studies for a masterplan let identify the degraded condition of the walled city and of its heritage, a physical degradation as well as an overall loss of identity due, in addition, to traffic congestion and the lack of green areas. Starting from these facts the aforesaid interventions as well as a series of projects were implemented not to strengthen the cultural identity of those goods belonging to a single part, but to promote the recovery of architectural heritage of both parts (Baldestone 2007).

The role that a shared material heritage plays as a possible peacemaker among different communities is proved by the Venetian walls that emblematically bind the two parts of the city and were chosen as the logo of both municipalities.

However, the most iconic part of Nicosia is in the Buffer Zone that, even now, is directly controlled by the United Nations and, more specifically, by United Nations Peacekeeping Force in Cyprus (UNFICYP). It deals with an area where, for almost sixty years, time seems to have stopped and hosts such an interesting architectural heritage now fallen into ruin, and represents the most important example of "difficult heritage" in the capital city of the island.

In the first years of the Twenty-first Century, thanks to the efforts of Greek-Cypriot and Greek-Turkish technicians, the masterplan undertook research in this area. The team identified degraded historical buildings inside the Buffer Zone and safeguarded more than 200 of them. In addition to a significant quantity of digitalized data that may be helpful for future projects, experts clearly indicated the need for some temporary

Fig. 1 Saint Gregory church in Famagusta found in advanced material decay and in need of structural emergency intervention for its consolidation and further preservation works. Photo by the Authors, 2021

works and emergency interventions for structural safety of the facades that were necessary to avoid possible demolitions as well as urgent restoration works.

In 2011, this study, whose symbolic value is also a source of great interest, was awarded by Europa Nostra as a fundamental element to preserve the heritage of the city center, as this area and its heritage will play a key role for "the unification and integration process of the city" (Europa Nostra 2011).

4 Perspectives on Cultural Heritage as a Peacemaker

As we tried to outline, the nature of cultural heritage and its meaning for the communities it represents, is crucial in conflict and crisis phases (Ristic and Frank 2019). Indeed, it can simultaneously become a tool and final purpose of a crisis. It is a tool as an instrumental cause to trigger conflicts based on the values it represents; on the other hand, it is the purpose, because, after its destruction, not necessarily intentional, the memory and history of a population are destroyed. In these cases, architectural, religious, and civil heritage becomes a symbolic element to undermine the basis of a population's identity, whose effects, as we have observed in the case of Cyprus, go

Fig. 2 The 'ghost town' of Varosha (TRNC); a built cultural heritage created by the conflict and the division itself, which, only recently, started to be enhanced through touristic activities raising nonetheless, further ethnic debates and conflicts. Photos by the Authors, 2021

beyond the conflict itself. Furthermore, in peace periods, it has the power to deepen the memory of the split past. Thus, it becomes evident that cultural heritage plays a key role in the construction of a post-war memory and, as a result, considering the narrative that builds all around it, can contribute to the construction of a peaceful and shared future (European Union, European External Action Service 2021). As a result, considering the way politics, civil society, and the world of culture narrate it, it can become a key element for the construction of a peaceful future or, on the other hand, maintain and perpetuate the divide. "Yet, at the same time, we can see numerous examples of how engagement with cultural heritage can be a strong component in fostering social cohesion, through processes that acknowledge minorities, promote gender equality and actively seek to address structural inequalities and ongoing insecurities. Cultural heritage can be a site for community engagements, activism, a space for dialogue as well as claiming rights and acknowledgement bymarginalized groups" (Kappler and Mannergren Selimovic 2021).

Under this point of view, there are several projects and programs financed by the European Union, whose main purpose is to protect and preserve cultural heritage, being aware of its importance as an essential tool to strengthen the role that Europe plays as a peacemaker as well as to prevent conflicts. Of course, it is necessary to have a series of tools that outline its role in terms of social cohesion and minimize any possible element of division to transform cultural heritage into a driving force for peaceful development of a society. From this point of view, specific political choices at local, national and international levels are necessary. As cultural heritage, and, more specifically, architectural heritage, is often cross-border and the result of multiple stratification processes concerning the history of different societies, it may be interpreted as a key factor in political choices. In case of international funding, as it usually occurs in crisis areas, the choice to promote preservation, conservation, and enhancement interventions of that heritage that symbolizes cohesion instead of division, as it is shared by different communities, may be a key driving force. At the same time, international awards and recognitions may represent a pull factor in this direction, if their object is represented by projects that are not only successful, but also carried out by teams composed of experts from different communities with reference to objects and areas of a shared heritage, for instance the 2011 award *Europa Nostra*, dedicated to study projects in the Buffer Zone of Nicosia.

As regards Cyprus, many attempts have already been made. The bi-communal Nicosia Masterplan, restoration works of churches and mosques in both parts of the island, the Association for Historical Dialogue and Research (AHDR), founded in 2003 and the Home for Cooperation are the proofs of these attempts. This last one, whose seat is a building dating back to the 1950s of the Twentieth Century in the UN Buffer Zone of the capital city, was recovered, as it was part of a group of degraded and abandoned buildings in that area. Now it is possible to get access to it from both parts of the city. Consistent with the *European Heritage Strategy for the 21st Century*'s goals, all of Home for Cooperation's activities are focused on education and promotion of intercultural dialogue, considering the key role of cultural heritage as a possible driving force.

Fig. 3 Built cultural heritage within Nicosia Buffer Zone. Photos by the Authors, 2021

Making people aware of the role that cultural heritage may play as a peacemaker as well as the promotion of tolerance are key elements to mitigate the divisive effect it may have in conflict and crisis areas. Furthermore, education of the youngest generations is a crucial factor, from primary schools to universities, but it also involves all those institutions that, at local level, can contribute to highlight common heritage or, anyway, the sense of respect for the memory of other cultures. Indeed, interventions on cultural heritage should not impose an almost always temporary reconciliation depending on external factors, but, on the contrary, they should promote a harmonic respect of differences based on reality, physical, and social features of the place where this occurs. This should happen when it deals with a cultural heritage resulting from conflicts, i.e., memories, destroyed palaces, commemorative monuments dedicated to dead soldiers, and cemeteries that the war itself created. This is a cultural heritage that, more than the most common one, may be a division or a reconciliation factor (Fig. 2).

"More informal traces of the conflict become immersed in everyday life and are imperative for how meanings and narratives are constructed around notions of divisive pasts and possibilities for shared futures" (Kappler and Mannergren Selimovic 2021). This is a perfect description of Nicosia Buffer Zone, a daily memory of the forced division that has the potential to become a driving force of a possible reconciliation through architectural heritage (Fig. 3). Specific restoration and rehabilitation interventions on the many damaged historical buildings and their opening to public use from both parts of the city, as in the case of Home for Cooperation, can be an already experimented and multiplied example of using the existing cultural heritage for bi-communal dialogue, cooperation, and reconciliation.

References

https://www.osce.org/files/f/documents/3/e/522616.pdf

Balderstone S (2009) Cultural heritage and human rights in divided Cyprus, in Langfield M, Logan William, Craith Mairead N (eds), Cultural Diversity, Heritage and Human Rights. Routledge, London

Balderstone S (2007) Managing Heritage in the wake of War and Conflict in Cyprus. ICOMOS Conference Extreme Heritage, James Cook University, Queensland

Baroncini E (ed) (2019) Il diritto internazionale e la protezione del patrimonio culturale mondiale. Università di Bologna, Bologna

Carstens A-M, Varner E (2020) Intersections in international cultural heritage law. Oxford University Press, Oxford

Carter, N., Martin, S. (2019) Dealing with difficult heritage: Italy and the material legacies of Fascism, Modern Italy, 24(2): 117-122

Constantinou Costas M, Demetriou O, Mete H (2012) Conflicts and Uses of Cultural Heritage in Cyprus, in Journal of Balkan and Near Eastern Studies, 14(2): 177-198

Council of Europe (2005), Framework Convention on the Value of Cultural Heritage for Society, CETS NO. 199, Faro, 27.X.2005

Council of Europe (2021a), Concept on cultural heritage in conflicts and crises. A component for security and peace in European Uninon's external action. https://data.consilium.europa.eu/doc/document/ST-9962-2021a-INIT/en/pdf

Council of Europe (2021b), European Heritage Strategy for the 21st Century. https://www.coe.int/en/web/culture-and-heritage/strategy-21
Europa Nostra (2011) Architectural heritage of the buffer zone in the walled city of Nicosia. https://www.europeanheritageawards.eu/winners/architectural-heritage-buffer-zone-walled-city-nicosia/
European Union, European External Action Service (2021) Concept on Cultural heritage in conflicts and crises. A component for peace and security in European Union's external action. https://data.consilium.europa.eu/doc/document/ST-9962-2021-INIT/en/pdf
ICOMOS (2015) People-centred to the Conservation of Cultural Heritage: Living Heritage. https://www.iccrom.org/section/people-and-heritage/people-centred-approaches
Kappler S, Mannergren Selimovic J (2021) Working with the cultural heritage of conflict for peacebuilding: Lessons learned from the Western Balkans. Civil Society Dialogue Network Discussion Paper, 16, Bryssel, European Peacebuilding Liaison Office
Macdonald, S (2009) Difficult Heritage. Negotiating the Nazi past in Nuremberg and beyond. Routledge, London-New York
Maniscalco, F (2006a) Il patrimonio culturale in Medio Oriente fra Jihad, Intifada e "Guerra al terrorismo". http://www.webjournal.unior.it - (II) 2006a
Maniscalco, F (2006b) World Heritage and war. Linee guida per interventi a salvaguardia dei Beni Culturali nelle aree a rischio bellico. Massa, Napoli
OSCE (2022) Report on violations of international humanitarian and human rights law, war crimes and crimes against humanity committed in *Ukraine (*1 april – 25 june 2022).
O'Keefe R (2006) The protection of cultural heritage in armed conflicts. Cambridge University Press, Cambridge
Ristic M, Frank S (eds) (2019) Urban Heritage in Divided Cities: Contested Pasts. Routledge, London
Roversi Monaco G.M. (2019), L'evoluzione della tutela dei beni culturali in caso di conflitto armato: dalla Convezione dell'Aia del 1954 alla giurisprudenza penale internazionale, in Baroncini 2019: 27-50
Tunbridge J E, Ashworth G J (1996) Dissonant Heritage: The Management of the Past as a Resource in Conflict. Belhaven Press
Verdiani G, Camiz A (2020) Fatto senza mani: rilievo, storia e valorizzazione del monastero di Panagia Acheiropoietos, Cipro, Firenze Architettura 24: 84–89

Engaging Impossibilism in Planning Practice. Reflections from the Case of Nicosia

Luca Gaeta and Gabriele Pasqui

Abstract This chapter applies the concept of impossibilism (initially developed by Marxist political theorists) to make sense of the bi-communal planning practices undertaken in a city structurally divided along lines of ethnicity and military occupation. The Nicosia Masterplan is a four-decade-long bi-communal initiative to comprehensively plan the city's future, whether political reconciliation will be achieved or never. By pushing the notion of impossibilism beyond the limits of reform under capitalism—as Marxists have it—we aim to frame a planning process entailing a radical degree of political uncertainty as part of its mandate. In doing so, we are mindful but also critical of Hirschman's possibilism as insufficiently accounting for the Cypriot planners' experience of active disillusionment. Impossibilism describes a condition of political stalemate in which no realistic solution exists for the main problems of urban coexistence. However, planning actions are necessary to make everyday life tolerable. The conclusions discuss tactics and tools for enabling spatial planning when cities are suspended in a state of impossibilism with long-term socio-spatial imbalances.

1 Introduction

How to practice spatial planning in Nicosia, the divided capital of Cyprus? What are the reasonable conditions for planning in a city structurally divided since 1974 along lines of ethnicity and military occupation?

This chapter aims to reflect on planning opportunities and constraints in a condition of political stalemate, in which no realistic solution exists for the main problems of urban coexistence. From this perspective, we use Nicosia as an example of cities

L. Gaeta (✉) · G. Pasqui
Department of Architecture and Urban Studies, Politecnico di Milano, Milan, Italy
e-mail: luca.gaeta@polimi.it

G. Pasqui
e-mail: gabriele.pasqui@polimi.it

© The Author(s), under exclusive license to Springer Nature Switzerland AG 2023
A. Buoli and O. C. Ţiganea (eds.), *Territorial Fragilities in Cyprus*, Research for Development, https://doi.org/10.1007/978-3-031-36076-3_6

and territories characterized by severe conditions of radical uncertainty and long-term conflicts. In such contexts, planning tools and strategies have no chance to cope with deeply distressing socio-spatial issues and effectively transform everyday life, thus improving cities.

By adopting a concept initially developed by Marxist theorists, we have decided to use the word "impossibilism" to describe situations in which long-term socio-spatial imbalances seem to prevent any transformative planning action claiming real change (Albrechts 2010). As we shall explain, impossibilism tries to cover situations in which the effectiveness of public action is suspended and attempts to generate comprehensive solutions turn out illusory.

The chapter is organized into five sections. After the introduction, Sect. 2 presents and discusses the case of Nicosia Masterplan. This is a four-decade-long bi-communal initiative to comprehensively plan the city's future, whether political reconciliation will be achieved or never. It can be considered an attempt to produce a shared vision and specific solutions for Nicosia crossing the physical and symbolic boundaries that divide the city. The troubled experience of Nicosia Masterplan is reconsidered in Sect. 3 under the notion of impossibilism, discussed in a theoretical perspective with reference to planning and social sciences debates. In Sect. 4, impossibilism is compared to Hirschman's possibilism as insufficiently accounting for the Cypriot planners' attitude of active disillusionment. In the conclusive section, indications for planning practice under conditions of impossibilism are given in a perspective sensitive to the everyday life experience of citizens and city users.

2 The Nicosia Masterplan: An Exercise of Impossibilism in Urban Planning

Nicosia Masterplan is the name given to a bi-communal urban planning initiative being conducted since 1981 in the capital city of Cyprus. In the aftermath of the events that led to the occupation of North Cyprus by Turkish troops, Nicosia experienced the burdensome condition of a divided city (Calame and Charlesworth 2011). Cut in the middle with a strip of no man's land, with no crossing allowed, the city faced the interruption of roads and basic infrastructures, the cleavage of religious and ethnic groups in the populace, and the lack of a unified administrative governing body. A city that had functioned as a whole for centuries was forcefully split into separate halves.

The two mayors of divided Nicosia soon started confidential discussions to make it possible for the city to function despite the unresolved political crisis. An integrated sewage system was completed in 1980 as an outcome of this pragmatic effort (Demetriades 1998). The mayors worried about the uncoordinated urban growth resulting in sprawl, poor infrastructure and insufficient public services. Moreover, the Buffer Zone harmed the historic center and other contiguous areas on both sides with abandonment, decline in business activities, building dilapidation and feelings of insecurity among residents.

In this context, with no prospect of a political solution, the Nicosia Masterplan was started with the aim of preparing a planning strategy for the city. A bi-communal multidisciplinary team tasked with planning, designing and implementing the urban strategy worked jointly, albeit not physically in the same office space. The UNDP provided the team manager and hired international consultants from countries accepted by both communities. The rationale was to prepare a master plan adaptable to a broad spectrum of future circumstances. More precisely, the master plan had to be made in such a way as to work either if reunification would be achieved in the years ahead or never. So, since the beginning, this informal bi-communal planning initiative has integrated considerable uncertainty as part of its conceptual and operational mandate.

The planning team in the early years was committed to developing a comprehensive urban strategy for greater Nicosia. This task not only entailed flexibility and adaptation, as every master plan does. It confronted the reality of two urban sections developing along rapidly diverging lines. Southern Nicosia was suburbanizing, and foreign capital promoted industrial investments. Northern Nicosia was undermined by international isolation and the influx of poor migrants from mainland Turkey, for the most part not integrated with more secular Turkish Cypriots (Damgacı and Dağlı 2018).

Since 1985, the master plan's focus was shifted from the metropolitan area to the urban core with the aim of preparing a detailed operational plan for the walled city. Surrounded by well-preserved walls built in the sixteenth century by the Venetian rulers, the oldest part of the city could be considered a common heritage for both communities. Its commercial area surrounding Ermou street, once the place of most intense intercommunal gathering (Bakshi 2015), lay entirely within the no man's strip. Physical decay and socioeconomic decline reinforced each other in the adjoining areas, thus tangibly affecting urban life. Amongst the trickiest problems to be tackled was the uncertain legal status of the properties abandoned by owners who fled after the Turkish intervention, with the side effect of preventing action by the planning authorities.

The master plan leveraged on the value of common heritage to encourage the flow of new residents and businesses. This goal was pursued by restoring civilian and religious monuments, replacing dead-end streets and alleys with new one-way circuits for motorized and pedestrian traffic, listing heritage buildings in the Buffer Zone, and improving public space to revitalize shopping areas. The planning team produced and displayed a unified map of the walled city on street corners, showing the parts each community used to (and still do) cancel from official maps.

A series of regeneration projects was completed in 2002 in selected areas such as Chrysaliniotissa and Arab Ahmet. Funded by USAID and UNHCR, these projects are located symmetrically to the Buffer Zone to reflect a political balance between the two communities in space. In 2003, a turning point was reached with the opening of a checkpoint at Ledra Palace followed in 2008 by a pedestrian crossing at the junction of Ledra and Kyrenia streets. Local people, and young generations for the

Fig. 1 A street view of Chrysaliniotissa neighborhood. Photo by Alice Buoli, 2023

very first time, could visit the other side of the city with mixed emotions of curiosity and animosity (Casaglia 2020) (Figs. 1 and 2).

The subdivision of Southern Nicosia into seven municipalities with autonomous planning powers made the concentration of urban development around the built-up area a task increasingly hard to achieve. Despite this, since 2010, the bi-communal team has managed to convey the regeneration strategy in the binding land-use plans being adopted by the local governments. Two action plans were commissioned to international consultants for areas immediately outside the walled city to expand the urban core. Detailed planning was done for a circular park along the moat and the restructuring of two squares opening at the extremes of Ledra and Kyrenia streets, all potentially bi-communal projects.

Two parallel area planning processes took place drawing from Nicosia Masterplan as the bridging strategy. However, the official planning maps were colored white by respective municipalities on the other side of the Buffer Zone. More importantly, the accession of Cyprus to the European Union in 2004 had material impact on the implementation of the action plans. While European structural funds were available in the southern part of the island, the *acquis communautaire* was suspended in the occupied part, thus slowing the pace of urban regeneration in Northern Nicosia.

The conservative turn in Turkey led by President Erdoğan together with the repeated financial crises affecting the banking system in Cyprus has harmed the reconciliation process. The Nicosia Masterplan still faces the radical uncertainty of a

Fig. 2 A street view of Arabahmet neighborhood. Photo by Oana Cristina Ţiganea, 2021

double scenario with the Buffer Zone in place or without it. The decline of the historic center is not reversed significantly, and the gap is widening between the two parts of the city both socially and physically. Recent projects such as the new orthodox cathedral built near Omeriye mosque warp the concept of symmetrical development put forth by the Nicosia Masterplan in a genuine spirit of reconciliation. No less disruptive are projects taking place in the north side to make Turkish assimilation an irreversible process.

We contend that the concept of impossibilism provides a useful tool to make sense of a bi-communal planning practice lasting forty years in the context of geopolitical instability. Impossibilism is a Marxist doctrine formulated in the early twentieth century that denies any real prospect of overcoming capitalism through the path of economic and social reforms, instead urging revolution as the proper means (Coleman 1987). We subscribe to an alternative meaning of impossibilism in the planning domain. Our proposed meaning refers to any stalemate situation in which planning practice cannot tackle the main urban problems because of insurmountable geopolitical barriers. When this kind of situation occurs, the impossibility to accomplish the desired objective is embedded in the action frame. The driving force of planning then is not the possibility to achieve a better future but rather the experience of an active disillusionment. Therefore, this modified concept of impossibilism accounts for intractable and extreme conflicts undermining the transformative potential of planning practice. The action needed is primarily to ease the everyday life of people and make it more tolerable as with the case of divided Nicosia. In cities where planning cannot be used to solve entrenched conflicts and disputes, it inevitably transforms into action for a liveable present rather than a brighter future.

Clearly, this kind of remedial planning may originate from visions of intercommunal reconciliation and be thought of as part of a wider peace-building process. Impossibilism in Nicosia is not realpolitik nor the acceptance of the status quo as unchangeable. It is, instead, the experiencing of a lived condition, in which discomfort and resoluteness blend together into resilient wisdom that consultants from abroad can hardly feel the same way as local planners do.

Abu-Orf (2005, 2011) has empirically researched the interaction between local planners and international consultants in cities, including Nicosia, affected by deep political and spatial divisions. In those cases, he argues, the model of collaborative planning (Healey 1997), with its emphasis on the social construction of meanings through discursive practice, does not reflect what is observed in divided cities. Abu-Orf (2005: 55) points out that "planning takes place in a political context in which the agendas of choice making are selectively managed" according to specific power inequalities. He also remarks on a concern with purely technical problems rather than deliberation as "the knowledge systems of instrumental rationality appear to dominate planning practice" (Abu-Orf 2005: 52).

An additional motivation for focusing planning practices on purely technical problems is related, in our opinion, to impossibilism. Bi-communal action and negotiation on infrastructural and physical improvements can be remarkably less controversial and ideologically charged in the context of a divided city, where authorities do not officially recognize each other. The choice to tackle technical problems, albeit never politically neutral, consciously sets aside burning issues of military occupation, compensation for seized property, and remembered atrocities, with the emotional load that would make any progress extremely hard to imagine. At the same time, technical actions are not without effects on the areas which suffer the most in a divided city. Much can be done with limited resources to lessen intercommunal tensions by working on 'trading zones' (Balducci and Mäntysalo 2015) evenly distributed in space to sustain daily encounters and routines. In so doing, planning fosters the right to a decent daily life for people caught in a geopolitical fracture rather than a way out of conflict.

3 Impossibilism: A Tentative Definition

In twentieth century philosophical and political discussions on the limits and possibilities of planning, many criticisms have emerged against not only the effectiveness but the possibility of comprehensive socioeconomic and spatial planning.

Friedrich von Hayek, for example, emphasizes that centralized planning certainly cannot plan for future knowledge developments, but neither can centralize an immense amount of knowledge about specific circumstances of time and place (von Hayek 1948). More recently, Stefano Moroni, inspired by von Hayek, has proposed a radical critique of comprehensive spatial planning (Moroni and Chiffi 2021; Moroni and Cozzolino 2019). In Moroni's perspective, comprehensive spatial planning is not only impossible due to human cognitive limitations. It is also a subversion of the idea of law, imagining that law should be primarily considered a tool or means for achieving any desired end. This approach, according to Moroni, should be substituted by a form of (local) government in which "only 'framework-instruments' are used to regulate private actions (regarding plots of land and buildings), whereas patterning-instruments are employed solely as means to discipline and guide public actions (for instance, to provide public infrastructures on public land using public funds). On this view, the law is interpreted not as an instrument to achieve any sort of specific outcome, but rather as an abstract and stable meta-framework for the peaceful coexistence of many different individuals with different and continuously changing aims" (Moroni 2018: 305).

Beyond von Hayek's criticism, the reflection on the unexpected effects of social action (Boudon 1984) and the US debate on urban planning in the 1960 and 1970s—from Melvin Webber (Rittel and Webber 1973) to the radical positions of Paul Davidoff (1965) and onwards—have shed light on the technical and political limits of comprehensive planning, partly due to its technocratic nature and faith in the power of technology and computing, but also due to its intrinsically ideological dimension.

Our use of the concept of impossibilism differs from these debunking critiques to comprehensive planning and to the very possibility of spatial planning. The impossibility, which we suggest observing the case of Nicosia, is not of a logical nature. Nor is it strictly linked to the impossibility of managing and centralizing the necessary knowledge for planning. Moreover, impossible situations like that of Nicosia are not only characterized by wicked problems, even if some of these problems haunt impossible planning processes. As Horst Rittel and Melvin Webber have underlined, a *wicked problem* is difficult or impossible to solve because of incomplete, contradictory, and changing requirements that are often difficult to recognize. "Wicked" denotes resistance to resolution and refers to a problem that neither can be solved nor adequately stated.

Impossibilism means something different from the presence of wicked problems because it involves more than ambiguities and uncertainties in the definition of problems, as the cognitive, institutional and political obstacles that prevent plans from manifesting their effects.

For our purposes, and drawing from Nicosia's situation, we define impossibilism in spatial planning as a situation characterized by five aspects: history, visibility, uncertainty, mistrust and symbolism.

The first aspect is the historical *persistence of socio-political instability*. In contexts such as Nicosia, planning is impossible for historical reasons linked to unresolved conflicts and persistent conditions of political instability. A situation becomes impossible throughout the long duration of history when problems become so rooted in the life of a city that they are assumed as "natural" conditions. Political instability, coupled with the perception of a permanent state of crisis, is the common situation in divided cities as Chiodelli (2017) underlines with regard to Jerusalem.

The second aspect is the *limited visibility* of spatial planning issues in the context of strong political and ideological conflicts. Although the daily life of women and men in a divided city is profoundly influenced by the effectiveness of spatial planning, at least in terms of service provision and land-use organization, the distance of political and diplomatic logic from the demands of life puts the problems of spatial planning on the sidelines. In these situations, spatial planning is often considered a secondary issue, compared to socio-political conflicts, and it can easily be taken out of the agenda. Entering and leaving the priority agenda is one of the major problems for long-term spatial planning in a highly conflictual context: this seems to be the case for Nicosia.

The third aspect is the condition of *radical uncertainty* in which the attempt to design spatial planning strategies takes place. The word 'uncertain' means that there is something, be it a judgment or an event, in which we cannot 'distinguish' what is true from what is false, or also what will (truly) occur from what will not (truly) occur. Uncertainty, in this sense, refers to a lack of knowledge and, more specifically, to situations in which we know the type of consequences but cannot meaningfully attribute probabilities to the occurrence of the event entailing such consequences (Chiffi and Chiodo 2020). Moreover, an uncertainty is radical, or ontological, if it is not possible to reduce it to risk evaluation and management (Pasqui 2022). Accordingly, an impossible situation is a condition in which spatial planning cannot imagine reasonable futures due to radical (political) uncertainty (Abbott 1988).

The fourth aspect is the presence of a permanent condition of *mutual distrust and disappointment*. As Hirschman highlighted in *Shifting Involvements*, there are periods in which citizens, but also institutions, simply lose hope, thus making it much more difficult to activate those processes of bias for hope and hiding hand which are fundamental for the success of plans (Hirschman 1967, 1971, 1982). Spatial planning feeds on trust, because its effects depend on the way in which citizens, businesses and institutions use their indications and tools. In the absence of trust, plans are often simply abandoned and left aside.

The last aspect that characterizes a condition of impossibility seems to be the *assemblage of material and structural conditions with narratives and symbols*. As shown in the previous section, each material aspect that characterizes the condition of Nicosia as a divided city (for example, the economic divide between the citizens living in the two parts of the city or the uncertainty of the land and property ownership regime) has a very strong symbolic connotation. They are captured by discourses, both common, institutional and technical, which reproduce the divisions, building a collective imagination of othering (Paasi 2021) which is very difficult to change.

Considering these five facets of impossibility together, we see that impossibilism has historical roots, it is connected to discourses and imaginaries as well as to knowledge production, control, and management, it has to do with the instability of contextual conditions and with the changing nature of the public agenda.

As the description of the troubled process of design and partially aborted implementation of Nicosia Masterplan has demonstrated, there are situations in which the convergence of different factors, among which the persistence of conflict for decades, the socio-political instability of the context, and the radical uncertainty seem to condemn spatial planning to a condition of paralysis and impossibility.

What can planners do when facing a situation of impossibility? What connections does the notion of impossibilism we have introduced have with Hirschman's notion of possibilism and with the Marxist notion of impossibilism as a rebuttal of reform?

4 Possibilism and Impossibilism

In the introduction of *A Bias for Hope* Hirschman summarizes his notion of possibilism (Hirschman 1971). He starts from the recognition of the importance of variables and of the political dimension for the understanding of economic phenomena and observes that the field in which the intertwining between economic and political dimensions (as well as between interests and passions) is most evident is that of social change.

In social change, understood as a complex process of a cultural, political, economic and symbolic nature, Hirschman sees the clearest exemplification of an approach to the social world that highlights the unique rather than the general, the unexpected rather than the expected, the possible rather than the probable. In turn, this approach is based on a fundamental shift from intention to action, from design to implementation, from assumptions to consequences.

Hirschman does not advance a general theory of change: he does instead recognize a set of specific, circumscribed, local changes that can generate new beliefs, attitudes and values through action and the sensemaking process connected to it. For this reason, the cognitive dimension of planning processes takes on a strong importance: through local action, planners can build conditions for altering the preferences and, in the long run, also the strategies of the actors, thus creating the platform for a possible change.

Possibilism, understood as openness to alternative sequences to the ordered ones and as an alternative to a conception of change as a unitary and fully intentional project, finally has its roots in the notion of unintended consequences of human action.

Hirschman's statements should not be hypostatized or interpreted in an ideological key. Immediately after having highlighted the generative possibilities of unintended change, he underlines two aspects. First, it is possible, and perhaps probable, that unintended change may be partially unsatisfactory for those who benefit from it. Furthermore, the two types of change are often closely intertwined. The change generated by unintended processes could therefore also pave the way for practices and processes based on voluntary change.

What matters, in Hirschman's perspective, is that it is possible, and desirable, to adopt a possibilistic logic if we want to help defend the right of an unplanned future. His passion for the possible is a posture to seize the opportunities hidden in the event of non-cumulative and unintentional processes, but also in the mutual interference between different forms of uncertainty.

As can be easily understood, Hirschman's possibilism is a sophisticated concept. It is not a general theory of planning, but a reflection on how to be part of a planning process. From this point of view, his lack of interest in reflecting about impossibility is clearly evident. However, impossibility represents a condition in which the minimal conditions to foster a bias for hope are not given, including a stable institutional framework and a convergence on the relevance of the problems addressed in the planning agenda. For this reason, the tactics and strategies for planning under conditions of impossibility should probably differ from those Hirschman set.

Our attempt to adapt the concept of impossibilism to planning practice under stalemate conditions is bound to be compared with Hirschman's possibilism as a symmetrical concept. Moreover, we shall acknowledge that impossibilism was elaborated initially by Marxist scholars to address the futility of reforms to end the structural exploitation of labor under capitalism. A further comparison with this initial conceptualization may help to clarify and differentiate our intentions.

Based on this Marxist doctrine, historically positioned in early twentieth-century Europe, an irreconcilable division exists in society between labor and capital that is spatially reflected in the urban fabric as the gap between center and periphery. Marxist impossibilism holds that reform-oriented action through gradualism and party politics cannot abolish class society altogether. Instead, reforms would integrate the working class into the capitalist system and give reactionary forces time to consolidate their supremacy. While, on the one hand, the efficacy of reform is denied, revolution is viewed as the only real possibility to defeat capitalism. Hence, the meaning of Marxist impossibilism limits itself to reform and entails the full possibility of revolution. But even more than being possible, the revolutionary event is prepared and awaited with almost messianic confidence in class struggle. Unconventional Marxist scholars as Lefebvre (2003) and Harvey (2019), despite the setbacks of the working class after the 1970s, weave together the ideas of revolution and the possible, claiming the city as a catalyst for struggle and self-organization. In their view, urban planning is strategically subservient to capitalist interests in the production of space. It works to reproduce the class division of society, with its spatial hierarchy, not to eliminate it.

In our revised conception of impossibilism, we too acknowledge the existence of a structural socio-spatial divide along lines of ethnicity and geopolitical othering with a heavy burden on the existence of the populations involved. Therefore, we undertake a different approach to make the concept meaningful for divided cities. In this context, we too doubt that spatial planning intended as social reform can bring reunification closer, but we do not do so by advocating insurgent forms of planning (Sandercock 1998). Instead, we acknowledge and value the determined effort of local planners to work on the threshold between two communities with all the uncertainty surrounding their plans. It all happens as if planning is not driven by prediction, yet it makes a difference in the ordinary life of communities in a divided city. Planners act vigorously under conditions of stalemate in a mood that could be described as active disillusionment. French intellectual Albert Camus has metaphorically captured this condition through the myth of Sisyphus, condemned by the gods to push a boulder uphill which always rolls back. This ostensibly meaningless task makes Sisyphus proud—according to Camus (1955)—as resistant to fate and, more importantly, attributing meaning to life under the most intolerable circumstances.

Impossibilism in divided cities refers to the unacceptable and continued separation of communities that used to share urban space and live together. In other words, the concept does not qualify planning practice as possible or not. Instead, it assumes that planning practice in impossible conditions without any clear future makes sense. Initiatives of bi-communal cooperation in a borderspace (Strüver 2018) are desirable and necessary when shared environmental resources and cultural heritage are at stake. Trust in the proximity of reconciliation may be a driving force to start cooperation after the drama of division. It happened in Nicosia in the early 1980s, when the Greek-Cypriot and Turkish-Cypriot mayors pragmatically met to envision the Nicosia Masterplan. Those men were familiar with each other, had a common professional background as lawyers, and could remember how life was before the division. In the long run, however, planning cooperation must base its continuation on renewed premises and mindsets. The expectation of reunification becomes less compelling, albeit not fading, and the division becomes ingrained in ordinary life. Thus, planning practice finds itself moving on a narrow path between the reality of partition on the one hand and the aspiration for bi-communal peaceful coexistence on the other. No complete breaking with the status quo of partition must be awaited to follow this path. Here is Hirschman's most helpful advice, to be complemented with a capacity to carry out planning and design without a bias for solutions which may never materialize. Under impossible conditions, like that of a divided city and island, every arrangement is stuck in suspension and temporariness. The Buffer Zone itself is nothing but a truce line indefinitely extended in time and highly bricolaged in space.

As Cypriot activist and researcher Mete Hatay makes clear, there is no such thing as a wholesale solution since "everybody [has] his or her own Cyprus problem" (interviewed by Horst 2022: 482). Hatay encourages observing ordinary people's interactions to realize how some spaces designed for everyday use "reconcile without any facilitation" (ibidem: 487), indirectly and ingeniously. From a planning perspective, it means creating urban spaces which allow bi-communal interactions as unintended outcomes. The stapling strategies and related projects presented in this volume pursue an improvement of interethnic relations through symmetrical place-making in the walled city and the reactivation of shared cultural heritage. As a further exercise of impossibilism conducted by an international group of architects and planners with no official mandate, it builds on Nicosia Masterplan, having learned its lessons (Figs. 3 and 4).

5 Conclusions

In this chapter, we have applied the concept of impossibilism to urban planning, and to do this, we have shifted its focus from methods to conditions. A divided city provides a fitting example of what we have referred to as impossible conditions. The

Fig. 3 The Samanbahçe Evleri neighborhood in north Nicosia. Photo by Oana Cristina Ţiganea, 2021

observations drawn from Nicosia and its bi-communal masterplan may be extended thus to other cities in which deeply rooted divisions are in place. However, it is essential to remark that the extension of the concept can be much broader and relate to conditions of radical uncertainty combined with socio-spatial inequality. From this viewpoint, dealing with impossibilism in urban planning might be more common than expected in times of multiple global crises.

It is clear to us that planning under conditions of impossibilism requires the preliminary disjuncture of means and ends. When the ends are shrouded in uncertainty and/or highly controversial, the selection of means cannot be based on them. Planners shall favor means according to criteria of bi-communal acceptability, spatial symmetry, essentiality for everyday life, and openness to multiple ends.

Urban life goes on despite impossible conditions, and planning can provide relief to inhabitants who legitimately aspire to lead a normal life. Rather than working for comprehensive strategies, planning is bound to reduce its ambition and act behind the veil of ignorance. Adopting this unpretentious attitude by no means prevents planners from grasping the moments in which tactical and incremental improvements can succeed without making progress towards a solution that may depend on exogenous geopolitical dynamics.

Tactical improvements are not to be confused with palliative measures. In impossible circumstances, questions of a general nature must be suspended without

Fig. 4 The new Nicosia municipality offices in south Nicosia. Photo by Alice Buoli, 2023

neglecting the disparity of resources between the conflicting communities as well as within them. Unlike Hirschman's hiding hand principle, which brings about creativity, we advocate hiding to mitigate the painful memory of tragedies still alive for many families in conflict situations, as in Cyprus. Tactical improvements may be seen as planning devices to open up opportunities for interaction and mutual tolerance in ordinary life, putting political tensions in the background for a while. Maybe planning cannot achieve strategic goals under impossible conditions, but it can do much to make life better.

References

Abbott J (1988) Understanding and Managing the Unknown. The Nature of Uncertainty in Planning. Journal of Planning Education and Research 24:237–251

Abu-Orf H (2005) Collaborative planning in practice: The Nicosia master plan. Planning Practice & Research 20(1):41–58. https://doi.org/10.1080/02697450500261707

Abu-Orf H (2011) Is planning possible in cities divided by violent conflict? International Development Planning Review 33(3):321–342

Albrechts L (2010) More of the same is not enough! How could strategic spatial planning be instrumental in dealing with the challenges ahead? Environment and Planning B 37:1115–1127. https://doi.org/10.1068/b36068

Bakshi A (2015) Nicosia Master Plan: Planning across the Divide. In: Björkdahl A, Strömbom L (eds) Divided Cities: Governing Diversity. Nordic Academic Press, Lund
Balducci A, Mäntysalo R (eds) (2015) Urban Planning as a Trading Zone. Springer, Cham
Boudon P (1984) La place du désordre. PUF, Paris
Calame J, Charlesworth E (2011) Divided Cities: Belfast, Beirut, Jerusalem, Mostar, and Nicosia. University of Pennsylvania Press, Philadelphia
Camus A (1955) The Myth of Sisyphus and Other Essays. Alfred Knopf, New York
Casaglia A (2020) Nicosia beyond partition: Complex geographies of the divided city. Unicopli, Milano
Chiffi D, Chiodo S (2020) Risk and Uncertainty. Foundational Issues. In: Balducci A, Chiffi D, Curci F (eds.) Risk and resilience. Springer Brief, Berlin-Milano, p 1–13
Chiodelli F (2017) Shaping Jerusalem: Spatial planning, politics and the conflict. Routledge, London
Coleman S (1987) Impossibilism. In: Rubel M., Crump J. (eds) Non-Market Socialism in the Nineteenth and Twentieth Centuries. Palgrave Macmillan, London
Damgacı AY, Dağlı UU (2018) Shifting Boundaries of Divided City Nicosia through Social Actors. Space and Culture 21(4):482–494
Davidoff J (1965) Advocacy and Pluralism in Planning. Journal of the American Institute of Planners 31(4): 331–338
Demetriades L (1998) The Nicosia Master Plan. Journal of Mediterranean Studies 8(2):169–176
Harvey D (2019) Rebel cities: From the Right to the City to the Urban Revolution. Verso, London
Hayek F von (1948) Individualism and Economic Order. University of Chicago Press, Chicago
Healey P (1997) Collaborative Planning Shaping Places in Fragmented Societies. Macmillan, Basingstoke
Hirschman AO (1971) A Bias for Hope: Essays on Development and Latin America. Yale University Press, New Haven
Hirschman AO (1982) Shifting Involvements. Private Interests and Public Actions. University Press, Princeton (NJ)
Hirschman AO (1967) Development Projects Observed. The Brooking Institutions, Washington (DC)
Interviewed by Cindy Horst (2022) Creating a Third Space in the Cyprus Conflict: Mete Hatay. In: Tønnesson S (ed) Lives in Peace Research. The Oslo Stories. Evidence-Based Approaches to Peace and Conflict Studies, vol 3. Springer Nature, Singapore.
Lefebvre H (2003) The Urban Revolution. University of Minnesota Press, Minneapolis
Moroni S (2018) Planning, Law, Ownership: Hayek and Beyond. Planning Theory, 17(2):305–310. https://doi.org/10.1177/1473095218767163
Moroni S, Chiffi D (2021) Complexity and Uncertainty: Implications for Urban Planning. In: Portugali J (ed) Handbook on Cities and Complexity. Edward Elgar, London, p 319–330
Moroni S, Cozzolino S (2019) Action and the City: Emergence, Complexity, Planning. Cities 90:41–52
Paasi A (2021) Problematizing 'Bordering, Ordering, and Othering' as Manifestations of Socio-Spatial Fetishism. Tijdschrift voor Economische en Sociale Geografie 112:18–25. https://doi.org/10.1111/tesg.12422
Pasqui G (2022) Coping with the Pandemic in Fragile Cities. Springer Brief, Milano-Berlin
Rittel HW, Webber M (1973) Dilemmas in a General Theory of Planning. Policy Sciences 4(2): 155–169
Sandercock L (1998) Towards Cosmopolis: Planning for Multicultural Cities. John Wiley and Sons, New York
Strüver A (2018) Europeanization in Cypriot Borderscapes: Experiencing the Green Line in Everyday Life, Geopolitics 25(3):609–632. https://doi.org/10.1080/14650045.2018.1550390

From No Man's Land to Everyone's Space. The Potential of Radical Design to Transform the *Green Line* in Cyprus

Laura Montedoro and Fabiano Lemes de Oliveira

Abstract The *Green Line* that crosses the island and splits the city of Nicosia in two is a wound that makes the geopolitical conflict that has been taking place in Cyprus since 1974 palpable. The military garrisons on both sides, the checkpoints, the UN trucks guarding what was once the main street of the city, and the ruins that overlook it offer an alienating and painful experience in the heart of today's Europe. Faced with the complexity of this situation, one wonders about the real potential of spatial transformation projects. How could an intervention in the physical space of the Green Line help foster dialogue between the two parties? And, more broadly, how could it be re-signified from nobody's land to everyone's space? The operability and effectiveness of design (landscape, urban and architectural) in a situation of long-standing socio-political conflict are deeply challenged; the spaces for interventions capable of making sense are indeed limited. In other words, it is a question of accepting the conditions of a minority in which architects and urban planners find themselves acting when faced with the problems of such contexts; and, starting from a position of humbleness but also mobilizing the power of imagination. It must be recognized that operations of space re-signification are possible and can play a meaningful role in offering "neutral" and hospitable places for dialogue. Where there is a contested memory of places, the production of an otherness, of a radical rewriting, can contribute to constructing an inclusive and ecumenical space. Starting from the analogy between language and space, this chapter proposes a reinterpretation of the Green Line as a Mediterranean free space, and provocative design actions are hypothesized to reinforce its "pidgin" character as a condition for potential new beginnings.

L. Montedoro · F. L. de Oliveira (✉)
Department of Architecture and Urban Studies, Politecnico di Milano, Milan, Italy
e-mail: fabiano.lemes@polimi.it

L. Montedoro
e-mail: laura.montedoro@polimi.it

© The Author(s), under exclusive license to Springer Nature Switzerland AG 2023
A. Buoli and O. C. Ţiganea (eds.), *Territorial Fragilities in Cyprus*, Research for Development, https://doi.org/10.1007/978-3-031-36076-3_7

1 The Green Line: An Open Wound

That Line which in 1963 was drawn on a map in green pencil (hence the name by which we still know it today) by General Peter Yang, an officer of the British contingent on the island, to separate the Greek districts from the Turkish ones would become—after the inter-ethnic clashes and the 1974 events and conflict leading to the partition of the island—the demilitarized Buffer Zone operated by the UN: an area of about 350 km^2 which extends linearly for about 180 km, from Kato Pyrgos in the west to the British military base of Dhekelia in the east, with a small branch to the north, near Famagusta.

The Green Line crosses the city of Nicosia, Cyprus' capital city, splitting it in half, making the experience of this urban space alienating, painful and frustrating. The street network is continuously interrupted in the north–south direction by gates, metal fences, walls made of barrels, and barbed wire, when not by actual military-manned checkpoints. Similarly, as the Buffer Zone approaches, the building fabric appears increasingly decadent, to the point of constituting a veritable curtain (or city blocks) abandoned along the Green Line, often also damaged by warfare. Where the Buffer Zone runs today, even if not in perfect alignment along the entire stretch, it was once the main artery of the city on the east–west axis: a commercial, vital and representative road; in other words, a linear heart. To get an idea, one can walk the last stretch west, at the entrance from Ektoros and Ermou Street, to get a glimpse of the functioning of the road section. On the other hand, to have a more precise perception of the state of abandonment and the consistency of the uninhabited building heritage coinciding with the Buffer Zone, the film "Smuggling Hendrix" by Piperides (2019) offers an excellent survey of otherwise inaccessible places: a ghostly scenario, sinister and surreal in the center of a European capital city.

If the Green Line is, therefore, a spatial device of truce (it was in fact established following the peace negotiations of 1974), it is also an absurd "open wound" (Grichting et al. 2012) which, of course, divides and unites (Papadakis 2006) like all borders, but which also makes physically palpable the strength of a conflict that is still alive, however controlled and kept latent (Casaglia 2020). Shared memories weigh on this strip of land, as well as on the whole territory of the island, but distinct and divisive, politically oriented and distorted memories are nurtured and emphasized; a selective process, fueled by nationalist pride, which compresses the common heritage until it is relegated to oblivion (Papadakis 1993). Memory, when shared, can undoubtedly be a driving force for encounters and dialogue. However, when it is so strongly oriented, it can work in the opposite direction, becoming operational for maintaining the conflict and crossing generations through the delivery of polarized memories, and of resentments, indirect to those who were not protagonists or witnesses of the former events. In other words, a tradition of narration is built, where tradition is understood from its Latin origin of tradĕre, composed of tra- "beyond" and dăre "to give", which meant "to deliver, entrust, transmit". Hence the suggestion of producing a new tradition, a new memory, through the modification of space (Fig. 1).

Fig. 1 and Fig. 2 Along the Buffer Zone in Nicosia. Photo by Laura Montedoro, 2021

2 Analogy Between Languages and Places

"You don't inhabit a country, you inhabit a language. A homeland is this and nothing else", wrote Cioran (1987). The analogy between languages and homelands, languages and places, and words and spaces gain conspicuous attention during the Twentieth Century. Starting from the studies of linguistics and semiotics, between the scholars of verbal communication and those of space, academic exchanges on this analogy intensify. As Consonni (2013) stated, "we come into the world not only in a place, but also in a language". If the relationship between humans and language and humans and places can be described as coincident, this relationship is peculiarly identarian, therefore exclusive and by nature non-pluralistic.

Language, every language, is, therefore, simultaneously a space that unites and separates, endowed with its own irreducible singularity. This plurality and uniqueness of language mean that no discourse can become absolute, or total; instead, this same *aporia* condemns each speaking being to a partiality that does not allow for replication. If it is a medium to understand each other, then language is also an instrument of separation (just like space): "rien n'est intraduisible en un sens, mais en un autre sens tout est intraduisible, la traduction est un autre nom de l'impossible" (Derrida 1996: 103). Starting from the distance that language produces between identity and otherness and from the betrayal to meaning that every translation cannot escape, at the end of the nineteenth century, the most nationalist and colonial of centuries, the need to have a language "of no one and of all" intensifies. The opportunity to launch an artificial and universal language is discussed by some scholars who will then find a first operational proposal in the elaborations by the Polish Ludwik Lejzer Zamenhof, known as Esperanto, still the best known and most widespread "international auxiliary language". The intent was to promote peace and understanding between peoples, but at the same time to protect minor languages of very local significance, the use of which is threatened by the hegemony of the strongest nations in the global geopolitical arena. In this sense, the dream of an "other" language for everyone, of a linguistic earthly homeland, is the dream of a linguistic democracy, free from influences, powers and totalitarianism. It is a hypothesis for many scholars more convincing than other pursuits, such as that of a perfect European language (Eco 1993). With all different purposes, history had already known of similar experiments, although not extended on a planetary scale. The Mediterranean basin was one of them. Here, in fact, for a very long time, the speakers have used a lingua franca to understand each other, to trade and make agreements.

3 The Invention and Use of Lingua Franca in the Mediterranean

Although linguists are not unanimous on its periodization, it is now established that in the Mediterranean basin, a *pidgin language* known as *Sabir* (Mac Carthy-Varnier 1852) had been used for centuries, most of all in port cities and along the coast (Cifoletti 1989). The pidgin is defined as "a simple form of a language […] with a limited number of words, which are used together with words from a local language. It is used when people who do not speak the same language need to talk to each other".[1]

Born for commercial and diplomatic purposes, to facilitate traffic and exchanges on the docks between speakers of different languages, the lingua franca of the Mediterranean was not only a functional language, "of necessity", but has known such use and diffusion to make *Sabir* the oldest and longest-lived *pidgin language* known to exist. About its origins, some scholars trace its appearance back to the period of the Crusades; more recent research excludes an onset in the Middle Ages and places its beginnings instead during the sixteenth century, when Turkish influence extended to the North African coasts, and in particular Algeria, where a means of communication was needed both between Christians of different Romance languages, and between Christians on one side and Arabs and Turks on the other. This is the reason why Sabir *is often referred to as a Barbary* lingua franca, as it was used, developed and propagated from the lands of "Barberia", i.e., from the Ottoman regencies of Algiers, Tripoli and Tunis (Cifoletti 2004). The temporal stability of this linguistic phenomenon, which lasted for at least three centuries with certainty, and the extent of its use to all daily life cases, and not only to strict mercantile needs, allowed it to spread considerably. As for the lexicon, despite the promotion of *pidgin* by the Ottomans, only a few terms were derived from Turkish; most of the words were of Italian origin, especially Venetian and Genoese, a smaller part was taken from Spanish and the rest of the terms were borrowed from Arabic, Catalan, Greek, Occitan and Sicilian.

It has recently been observed that, from a historical and philological point of view, it would be improper (and ideological) to assume the Barbary lingua franca as a mixed and irenic language (Venier 2016), in a romantic and consolatory vision. Sabir was not born as a language *of* peace. Nonetheless, we underline here the value of symmetrical otherness between speakers and of attribution of belonging to a community of a higher order coinciding with the historical–geographical context of the Mediterranean: a plural, articulated, conflictual unity, but still a unity. By inhabiting the Barbary lingua franca, one was also living in a larger homeland in which Arabs, Turks and European Christians could easily understand each other. That same Mediterranean "is mentioned a lot, but it is not known how much it is truly understood and accepted for what it is" (Ivetic 2022: 7).

---

[1] Source: Oxford Learner's Dictionaries, https://www.oxfordlearnersdictionaries.com/definition/english/pidgin. Accessed on March 2023.

4 From Analogy to Metaphor: Free Space as a Lingua Franca

Language is a structure made up of differences. So is space, as a place of deferral and differentiation. Space is always spacing. Moreover, language (at least from structural linguistics onwards) can be understood as a system of differences. The Cypriot Green Line is a device of differentiation, of spacing, but for this very reason, it is a space open to continuous processes of re-signification. The Green Line would therefore lend itself to corresponding to the opportunity to produce a free space spatially. Yet, how could this need be treated in a transformation project? How is good disposition to engage in otherness possible when the situation of co-presence is characterized by what Pizzorno (1993) defines as "conflicts of recognition", in which the parties do not belong to the same system of relationships and enter the conflict by bringing into play non-negotiable identities (and at the same time aiming to be recognized), which are themselves constituted and/or reinforced in conflicts? In other words, it is a question of asking how to produce the space for diversity to live together, a "space in which there is the possibility of radical otherness and at the same time of infinite proximity, independently of the sharing of a value or a foundation" (Pasqui 2018: 21); or as Nancy (2001: 35) suggested, designing space for a "community without a common origin".

One possibility seems to us that of radical rewriting: where the memory of places is stratified—disputed, polarized, denied, removed, erased, but at the same time shouted by the absence of meaning of the space reduced to a demilitarized Buffer Zone. Perhaps only the production of a new imaginary, totally other and freed from memory, can contribute to constructing an inclusive and ecumenical space. A "*pidgin space*", which no longer belongs to one side or the other, not for security reasons, but to make dialogue possible. Returning to belonging/being above all Mediterranean, that belonging to the Mare Nostrum, which, albeit with an extraordinary plurality of recognizable and singular identities, has been typical of the peoples there for millennia. Another element that can be taken as common ground is insularity. As Matvejević reminds us, "islands are very special places [...]; the peculiarity common to most of them consists in waiting for what will happen" (Matvejević 2020: 30–31). The permanent condition of separation (from the mainland) becomes an ontological trait of their inhabitants, like their unique relationship with time. Insularity, which by its nature divides, unites or can unite the people of the islands.

Starting from these assumptions, radical architecture exercises show how re-signification devices can work. In fact, radical architecture was confronted with that question matured, for example, within the Team X concerning the rapidly changing society: "can architects respond to the social demand for plurality? (...) If society has no form, how can architects build its counter form?" (Van Eyck 1962). A question that is anything but trivial that focuses on one of the main challenges for the production of space in the contemporary world (Fig. 3).

Fig. 3 Along the Buffer Zone in Nicosia. Photo by Laura Montedoro, 2021

5 The *Green Line* as everyone's Space: The Potential of the Radical Project

The radical project is an opportunity to escape, to use Nietzsche's words, the grasp of the "ever increasing burden of the past" that can push "the human being down and bows him over" (Nietzsche 1874), and seeks that state found in a child that "does not yet have a past to deny and plays in blissful blindness between the fences of the past and the future" (*ibidem*). It suspends time, working on that borderline at which the past is prevented from freezing the present (Nietzsche 1874) or stalling the imagination of possible, potential, and preferable futures. In the history of the twentieth century, the so-called "radical architecture" put forward, already from the end of the fifties and peaking in the 1970s, innovative proposals that challenged the functionalism that had dominated the scene in the previous decades. The Japanese Metabolists, for instance, as well as the Italian experience (Mello 2017) sought to employ visionary thinking to mobilize the forces at hand, presenting new ways of relating humans with the environment and ultimately forging new territories of existence (Lemes de Oliveira 2011). Of the many declinations that radicalism has known, we seek to retrieve the utopian and visionary vein of their proposals. As such, we ask how radical thinking for the Green Line could be an instrument of transformative power. Some possible examples:

The Olive-Green Line: the Buffer Zone is transformed into a large olive grove. The choice falls on the olive tree as the quintessence of the Mediterranean spirit. For a long time, it was believed that it was precisely the distribution of the olive tree that identified the Mediterranean area. It could represent a higher order identity that contains all, the diffusion area of the *Sabir*. The olive tree is seen as a symbol of peace; it is a long-lived plant that requires care; it gives rise to a productive forest that bears fruit; the olive tree is the minimum unit of a familiar and common landscape. This approach sees the landscape project as a peace pedagogy project. Strengthened by participatory processes with the two parties, with the active involvement of children and young people for the preparation of the land, and the planting of young plants, in a collective ritual.

The More-than-human Green Line: Jacoby's (1999) claim that utopianism and universalism are necessarily attached makes us reflect on how its disruptive power may allow us to access new life possibilities and address global challenges from a transformative standpoint. Besides the geopolitical issues at play discussed previously, the climate and biodiversity crises challenge the continuation of life on the planet and pose the need for fundamental reconsiderations of our relationship with the biosphere. A more-than-human Green Line becomes a space where biodiversity is a vector for a reconnection to nature and ultimately, a point of convergence between peoples. It offers a multispecies (Celermajer et al. 2021) understanding of place and focuses attention on the intrinsic and relational values of nature, enhancing empathy and cooperation both between humans and between humans and other species (Clemént 2004). Language is expanded to encompass communication across the living world (Kohn 2013).

The Blue-Green Line: Nature is mobilized as the selected past for a shared future. This vision calls for the re-establishment of the river, evident in medieval maps, that was erased of the city's fabric with its diversion outside the city (Papadakis 2006). It emphasizes the possibility of using water as a symbol of flow, adaptability and transformation. The axis becomes an oasis against the predicted increase in urban temperatures due to climate change. As a mutable flooded space, it creates a third margin within the strip, also defining new porous edge conditions, and encouraging permeability and exchanges.

The Shelter Line: Aa place in which climate shelters serving both sides of the Line replace military infrastructure and provide refuge for people and non-human beings. A transformation of the territory that aims at antifragility and from a standpoint of the universal need for protection finds a common ground for dialogue.

We could go on, for example, trying to develop even more radical and provocative hypotheses, such as *The Red Green Line*: perhaps in a less ambitious but equally impactful way, the ground could be colored—as in an art installation—to build a large *common playground*; or we could imagine *The Floating Green Line*, even more provocatively, working with the idea as a sort of extrusion of the surface of the green Line, as perhaps Yona Friedman might have appreciated (Friedman 2003) or even *The Fake Green Line, a new venetian dream*, since Venice has dominated this part of the Mediterranean Sea for many years: what if the Green Line is transformed in a fake venetian landscape? Still today, a great part of historical heritage comes from the Venetian age, especially the great system of fortifications. A space no longer Greek, no longer Turkish, but the character of one of the hundred threads of the island's millennial history (Fig. 4).

Fig. 4 Along the Buffer Zone in Nicosia: A dead-end on the southern side of the city. Photo by Laura Montedoro, 2021

6 Conclusions

This chapter reflected on the possibilities of design in facing the challenges of transforming a space frozen in time by a long-standing socio-political standstill. It explored the relationships between language and space, with particular attention to how the principles of commonality seen in the idea of a lingua franca can be considered, first through analogy and then as a metaphor, in the envisioning of alternative futures as a free space for the Green Line. In so doing, the chapter puts forward the notion of a *pidgin space*, a new lingua franca/free space, to be co-constructed and co-lived. The potential of radical thinking in envisaging ways in which the notion of a pidgin space might be framed is explored. Radical design often suspends time and locates itself outside the constraints of technical, economic, and political feasibility. Its disruptive power has the potential to mobilize critical visions of what might become. In doing so, it has the potential to generate reflections, start discussions, re-ignite conversations towards mutual understanding and co-creation of a new preferable future.

References

Casaglia A (2020) Nicosia Beyond Partition. Complex Geographies of the Divided City. Unicopli, Milano

Celermajer D, Schlosberg D, Rickards L, Stewart-Harawira M, Thaler M, Tschakert P, Winter C (2021) Multispecies justice: theories, challenges, and a research agenda for environmental politics. Environmental Politics 30(1-2):119-140. https://doi.org/10.1080/09644016.2020.1827608

Cifoletti G (1989) La lingua franca mediterranea. Unipress, Padova

Cifoletti G (2004) La lingua franca barbaresca. Il Calamo, Roma

Cioran EM (2007) Confessioni e anatemi. Adelphi, Milano

Clemént, G (2004) Manifeste du tiers paysage. Éditions Sujet/Objet, Paris

Consonni G (2013) Da grande voglio fare il poeta. La vita felice, Milano

Derrida J (1996) Le monolinguisme de l'autre ou la prothèse d'origine. Galilée, Paris

Eco U (1993) La ricerca della lingua perfetta nella cultura europea. Laterza, Roma-Bari

Friedman Y (2003) Utopie realizzabili. Quodlibet, Macerata

Grichting A, Castrillo MC, Keszi S, Frangoudi G (2012) Stitching the Buffer Zone. Landscapes, Sounds and Trans-Experiences along the Cyprus Green Line. Nicosia, Bookworm Publication

Ivetic E (2022) Il grande racconto del Mediterraneo. Il Mulino, Bologna

Jacoby R (1999) *The End of Utopia: Politics and Culture in an Age of Apathy*. Basic Books, New York

Kohn E (2013) How forests think: toward and anthropology beyond the human. University of California Press, Berkley

Lemes de Oliveira F (2011) Of Metabolism: future cities for our contemporary world. Risco 14(2):77-81

Mac Carthy-Varnier (1852). *La langue Sabir*. In L'Algérien, journal des intérêts de l'Algérie, 11 mai; now in Cifoletti, G (2004), pp. 262-68.

Matvejevic P (2020) Breviario Mediterraneo. Garzanti, Milano

Mello P (2017) Neoavanguardie e controcultura a Firenze. Il movimento Radical e i protagonisti di un cambiamento storico internazionale. Angelo Pontecorboli Editore, Firenze

Nancy JL (2001) Essere singolare plurale. Einaudi, Torino
Nietzsche, F (1874) *On the use and abuse of history for life*
Papadakis Y (1993) The Politics of Memory and Forgetting. Journal of Mediterranean Studies 3(1):139-154
Papadakis Y (2006) Nicosia after 1960: A River, A Bridge and a Dead Zone. Global Media Journal: Mediterranean Edition 1:1-16
Pasqui G (2018) Spazio pubblico e società plurale. In: Montedoro L (ed) Lo spazio pubblico come palinsesto, Maggioli, Sant' Arcangelo di Romagna, p 17–24.
Pizzorno A (1993) *Come pensare il conflitto*, in *Le radici della politica assoluta*. Milano: Feltrinelli.
Van Eyck A (1962) *Alison Smithson*, edited by Team 10, *Primer 1953–62*, "Architectural Design", December, p. 564
Venier F (2016) L'invenzione del consenso: il caso della lingua franca. Rivista Italiana di Filosofia del Linguaggio 1: p 292-309

Stapling Strategies Around the Green Line in Nicosia's Old Town

Francesco Pasta, Dafni Riga, Federico Barbieri, Wenshan Chen, and Wei Lyu

Abstract This essay outlines a theoretical and methodological framework for the proposed intervention strategies for Nicosia's walled city center presented in this section of the volume. The goal of this process, which we refer to as "stapling", is to contribute to creating a platform for local collaborative activities across and beyond the border as a way of tackling issues of shared identity. The chapter is divided into three parts. We begin by providing a conceptual background, relating a range of subjects that are relevant to our research and intervention approach in the divided city of Nicosia, such as, e.g., the concept of a "thick" border, the ambivalent semiotics of urban space, the gap between conceived and experienced space, the materiality of everyday life, and the idea of cross-border engagement. Grounded in the theoretical framework and based on fieldwork conducted in Nicosia, the second section identifies three main themes, which can be considered as both challenges and opportunities that the proposed strategies aim to address and develop from different perspectives. These are: (1) bridging inter-communal planning with civil society; (2) re-activating under-used or abandoned spaces; and (3) building upon a shared socio-ecological system. Such macro-subjects constitute criticalities that present the potential for action and intervention. The third section introduces the methodological approach shared by the three strategies. We briefly outline the principles of community-based action planning, the key guidelines for intervention, and the community actors to be involved. Based on these elements, we discuss how the proposed planning strategies do not constitute a prescriptive masterplan, nor an inflexible vision. Rather, they compose

F. Pasta (✉) · D. Riga · F. Barbieri · W. Chen · W. Lyu
Department of Architecture and Urban Studies, Politecnico Di Milano, Milan, Italy
e-mail: francescomaria.pasta@polimi.it

D. Riga
e-mail: dafnitheodosia.riga@polimi.it

F. Barbieri
e-mail: federico.barbieri@polimi.it

W. Chen
e-mail: wenshan.chen@polimi.it

W. Lyu
e-mail: wei.lyu@polimi.it

© The Author(s), under exclusive license to Springer Nature Switzerland AG 2023

A. Buoli and O. C. Ţiganea (eds.), *Territorial Fragilities in Cyprus*, Research for Development, https://doi.org/10.1007/978-3-031-36076-3_8

an open framework for strategic and concerted action. They are complementary but not necessarily interdependent. We conclude by defining a "stapling strategy" as a process that builds upon and develops existing socio-spatial interlinkages between the two sides of Nicosia's historic city center, metaphorically "stapling" or "stitching" together spaces that used to be contiguous and continuous but are now severed by the Green Line. While grounded in the socio-political reality of Nicosia, this methodological approach can relate to other divided urban contexts.

1 Introduction

The projects presented in this book explore strategies to address issues of inter-communal urban life and heritage preservation in the Nicosia's walled city, both in the current state of partition, as well as in the context of a possible future scenario of reunification. By focusing our attention across the Green Line in the walled city, we propose "stapling strategies" to expand the common ground between the two sides.

Our research develops complementary strategies for co-production activities of events, design interventions, and planning proposals dealing with Nicosia's urban fragilities from different perspectives. This "stapling" process contributes to creating a platform for engagement and collaborative activities across and beyond the border, as a way of tackling issues of shared identity.

The aim of this strategic framework is to provide examples on how to activate connections across the Green Line by strengthening a twofold network of spaces and people, built through material upgrading, urban reactivation and civic collaboration on both sides of the border. To expand a citywide inter-communal network of associations, groups, institutions and individuals working towards shared urban regeneration and co-production processes, could be a way to set aside political bias and help reconcile relations across local communities and different social groups. This process could contribute to and plug into the Nicosia Masterplan, establishing connections between inter-communal planning and civil society at large, as a parallel process of envisioning a common urban future.

1.1 Conceptual Framework

In order to elucidate the key concepts that will be used as operative tools throughout this chapter, we start by illustrating the theoretical background relevant to our research and intervention approach in the divided city of Nicosia.

To understand the complex nature of the Green Line that is transecting the walled city, we started by synthetically exploring the scientific literature devoted to borders and their spatial and symbolic role in urban contexts.

In the past decades, borders have attracted increasing attention across a variety of disciplinary fields, from political geography to sociology, and from anthropology to urban studies and planning. From various perspectives, scholars analyzed the re-articulation of borders in our globalizing "borderless" world and the contextual evolution of static territorial frontiers into a complex and dynamic apparatus of control, surveillance, and discipline (Newman 2006; Mezzadra and Neilson 2012). In Cyprus and in Nicosia, however, the so-called "border" also constitutes a very concrete presence, physically shaping the urban space and the urban experience. In fact, the Green Line is not merely a line representing an international boundary but is indeed marked by a certain "thickness", thus creating a transitional space between the two sides of the city. This space can be characterized as a no-man's-land, as a void aiming to create distance between two communities in conflict (Fig. 1). The fact that the thickness of the Green Line is not constant inevitably creates disanalogous spatial relations between the two sides, sometimes favoring views on the other side, oftentimes resulting in blind dead ends.

The Green Line is of interest not only in spatial terms, but also, and perhaps more importantly, as a symbol that manifests the unresolved conflict between Greek and Turkish Cypriots and their respective national, socio-cultural and spatial identities (Strüver 2018: 3). From an anthropological point of view, this "thick border" fabricates metaphors of cultural and ethnic discontinuity. According to Thomas Nail, the border is not merely a physical barrier, but rather "an active process of bifurcation that does not simply divide once and for all, but continuously redirects flows of people and things across or away from itself" (Nail 2016: 4). Thus, building upon this argument, the border is there to divide, but to also allow and orient the possibility of crossing. Nevertheless, while physical proximity to the border may result in a more pronounced perception of these narratives, it is important to highlight that the effects of the border extend far beyond its immediate vicinities.

The existence of the Green Line per se exemplifies a duality in the city of Nicosia: the border defines two sides, two narratives with more than often conflicting memories and identities. This duality is translated into, but also questioned throughout urban space, since many sites and buildings in Nicosia are at the same time places of division, conflict and separation, but also of contact and exchange, evoking a shared identity (Papadakis 2006: 3). The memory of the partition lives and breathes on both sides of Nicosia and its interwoven meanings are inscribed into space, thus resulting in spatial interpretations of identity, memory, and conflict. Papadakis (2006: 3) points out how some spatial arrangements reflect the different views on the partition, highlighting that from the Greek side the Green Line looks makeshift (thus interpreting the partition as temporary), while on the Turkish side a wall has been built, demarcating the Buffer Zone as an ostensibly stable boundary (Fig. 2). During a walking tour in Nicosia, Papadakis also referred to the fact that the flags positioned close to the Green Line represent the stance of each side: the Greek Cypriot flags are made of fabric, subject to the changes of the wind, while the Turkish ones are made of metal and thus always visible. Such a remark underlines the fact that dual meanings of partition can be reflected across different scales in material and spatial configurations in Nicosia (see the introduction to this volume).

Fig. 1 Soldiers walking along the Green Line in the RoC. Photo by Dafni Riga, 2021

Our research examined the intangible origins of this duality and the way in which it is manifested spatially across the Green Line. By addressing the material urban dimension of immaterial meanings and narratives, as well as the role of a shared history, socio-ecological system, and material culture, the workshop and the proposals presented in this volume regard tangible and intangible heritage as an incentive for cooperation. Instead of hiding memories "under the carpet", our objective is to encourage new spatial narratives and memories, enclosed within the material historical space.

Our research explored the social and urban nature of the "bordering" process on both sides of the Green Line. We investigated how locals inhabit these spaces and what kinds of meanings they ascribe to them, yet without excluding the stance of social actors and stakeholders with regard to the border, place-making and place-crossing. While acknowledging the existence of the border and observing the effects it produces in urban space, we do not take it for granted. This means that our research questions the border by exploring its socio-political construction and engaging with alternative narratives that challenge the static territorial division between two allegedly exclusive national spaces and communities.

In this sense, we were interested in working in the physical and figurative "crack" between the official meanings attached to urban space in the state's hegemonic narrative, and the divergent interpretations and uses developed by people: that is, in the gap separating the conception of space from its experience (Lefebvre 1991). As it stands out in the case of divided Nicosia, space is in many ways instrumental to the materialization of authority and maintenance of domination. States invest efforts and

Fig. 2 The edge of the Green Line seen from the TRNC (top) and the RoC (bottom). Photos by Dafni Riga, 2021

resources into designing, ordering and structuring their own territory; yet, the very same space is subject to the complex dynamics of use, interpretation and signification carried out by its users (Dovey 1999; Sudjic 2005; Storey 2012; van Houtum and van Naerssen 2002: 125).

Deleuze and Guattari (1988) conceptualize striated and smooth space, whereby the striated is a space of stabilized, bounded and socially controlled identities, maintained through choreographed spatial practices, while the smooth is characterized by movement and instability that make possible the emergence of new identities and upsetting spatial practices. Smooth and striated space is one twofold entity, that exists only in mixture, in a continuous process of mutual re-constitution. Similarly, for De Certeau (1984: xv), "groups or individuals already caught in the nets of discipline" may re-appropriate space, subvert its official meaning and thus deflect—when not challenged—the power of the "maker" through "clandestine forms" of "dispersed, tactical and makeshift creativity". Focusing specifically on Nicosia, Papadakis employs De Certeau's significant distinction between "place" and "space", which informs different interpretations of space depending on the position and role of different individuals, collectivities, and institutions:

> De Certeau's approach places emphasis on praxis, on how social actors reinterpret, manipulate and tactically employ official constructions for their own ends. [...] 'Place' is what lies at the opposite of 'proper space' [...]. Proper space is constructed from above by officials and town planners with rationalist or political considerations. [...] Yet, local actors [...] give local meanings to 'proper spaces' as they walk through them, live in them and infuse them with their own memories and significations, in short, as they act through them (Papadakis 2006: 9).

Everyday life thus might be conceived as a lens through which to understand and explore the social and material practices and flows unfolding around the border, both in terms of tangible and intangible heritage, and with regards to current spatial practices and their potential to influence future narratives regarding the conflict. According to Navaro-Yashin, who examined the intertwined processes of construction of spaces and narratives in the Turkish Republic of Northern Cyprus (TRNC) from an anthropological perspective, "the imagination that goes into fabricating something is part and parcel here of the materiality of this manufacture, a process of making-and-believing, or believing-and making, at one and the same time" (2012: 6). In other words, beyond the opposition between social constructionist and materialist approaches, the material crafting of a territory and the imaginative creation of a social narrative need to be seen as a twofold process evolving in unison. Since borders are "socially produced and reproduced, and thus are always susceptible to being modified, transformed, erased, recreated, reimagined, transgressed" (Soja 2005: 34), our aim was to develop strategies focusing on material spaces along and across the Green Line, which could potentially function as generative anchors for social innovation by tackling issues of deprivation and isolation (Gaeta 2021).

In this light, the concept of "borderscaping" emerges as a useful methodological instrument for interpreting spaces on the edges of the Green Line and in the entire walled city, and a critical design tool for potential spatial interventions. As an approach, borderscaping refers to a civic border-writing process and a design method, enabling the emergence of shared practices of belonging, living, surviving, and resisting around borders (Brambilla 2015; Van Houtum and Eker 2015). Conceptually, it interprets borders both as dense "deposits" of historical layers and, at the same time, as sites for experimenting alternatives and future scenarios (Buoli 2020). We thus regard the border as a transcalar support and a spatial pivotal element that allows us to imagine future configurations of "cooperative" interaction on relevant social and urban issues. The research projects presented in this book, by proposing thematic interventions on selected sites, explore the potential of borderscapes to generate shared practices of living (Buoli 2020), which could ultimately contribute to renegotiating and reshaping the border itself.

In order to encourage and build upon existing cooperation efforts between the two communities, it is necessary to enhance the feeling of mutual engagement. According to Ker-Lindsay (2019: 27), the notion of "engagement without recognition" carries the assumption that interaction is better than isolation, but to produce results, there needs to be an underlying culture of engagement: some key strategies for achieving this are building ties and trust (breaking the culture of suspicion); finding joint solutions to common problems; and coordinating activities. Among the fundamental steps to enhance interaction across communities in Cyprus, Ewers (2018: 14) underlines the crucial role of opening more crossing points across the Green Line, and particularly the Ledra-Lokmacı pedestrian crossing in 2008, in dismantling the physical limitations between the two communities and opening up new opportunities for future collaborative development. At the same time, under the scope of borderscaping, it appears crucial to integrate bottom-up forms of "border disobedience" into planning regulation systems and standards in order to bypass institutional planning "rigidities" (Buoli 2020). Following these arguments, our aim is to suggest ways in which the urban borderland surrounding the Green Line within the walled city will become lively again as areas characterized by inter-communal engagement and interaction (Fig. 3).

Co-production may be defined as a process of strengthening participation and creative engagement among citizens, by involving them in the design, production and delivery of a project that will meet their needs. "Co-production means delivering public services in an equal and reciprocal relationship between professionals, people using services, their families and their neighbours. Where activities are co-produced in this way, both services and neighborhoods become far more effective agents of change" (Boyle and Harris 2009: 6). The goal of this process is to create a feeling of mutual support and shared responsibility between public actors and citizens. The Nicosia Master Plan is considered the first and most important joint planning effort (although not of participatory nature) carried out in Cyprus after the partition. According to Ewers (2018: iv), rehabilitation of decayed structures on both sides of the Buffer Zone promoted peaceful interaction, local economic growth, and resettlement in the historic core (Fig. 4), which led to population growth, the development of heritage tourism, and the setting of bicommunal cultural events.

The projects outlined in this book explore whether this can work on the scale of the walled city, with three different but interrelated thematic and spatial focuses. All proposals envision co-production processes encouraging the interaction among local actors and citizens. In our vision, this process of inter-communal socio-spatial engagement could eventually be conducive to the potential "opening" of the border - or even, someday, to its dismantling.

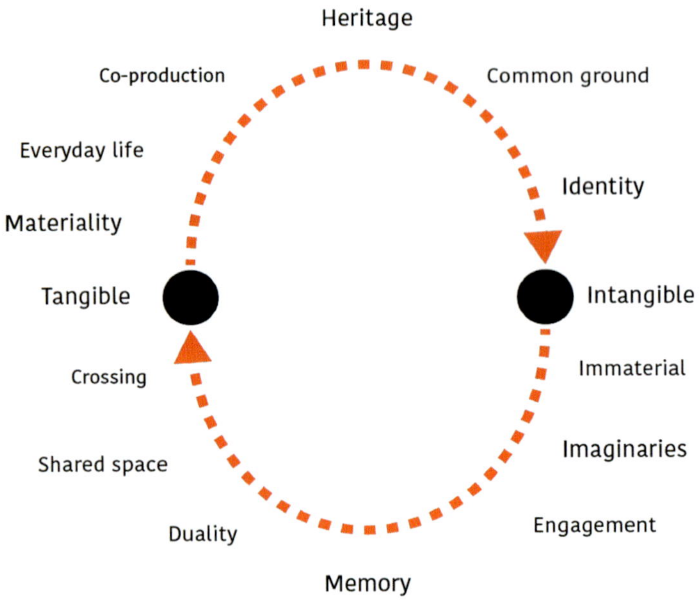

Fig. 3 Conceptual keywords. Elaboration by the authors

Fig. 4 Houses on the edge of the Green Line, TRNC. Photo by Francesco Pasta, 2021

2 Issues, Topics and Proposed Strategies

2.1 *Challenges and Opportunities*

Grounded in the theoretical framework previously outlined, we propose intervention strategies for Nicosia's walled city, which address, through different approaches, three main themes. The three macro-subjects discussed below can be framed as both challenges and opportunities. They constitute criticalities that present the potential for action and intervention, which the proposed projects intend to address and develop.

2.1.1 Bridging Inter-Communal Planning with Civil Society

By the term "inter-communal planning", we refer to planning processes that are jointly designed and implemented by a variety of entities across the Green Line divide in Nicosia's historic center and the city at large. Among such entities, the Nicosia Master Plan (NMP)[1] and the Technical Committee for Cultural Heritage[2] stand out. Moreover, across both sides, there are professional organizations, civil society associations, and informal groups which, although not working systematically in concert, carry out analogous projects and share a similar ethos and approaches. Our argument is that expanding the scope of inter-communal planning by bridging it with such existing realities could positively impact Nicosia's joint urban development. Therefore, in our research and fieldwork, we focused our interest on mapping out stakeholders active on both sides, exploring the effectiveness of cross-border collaboration among them, and proposing ways in which it could be scaled up.

We consider the NMP to be the cornerstone of shared strategies among the two sides, since it constitutes an extremely valuable framework for inter-communal cooperation and planning. Over the years, the NMP concretized several joint projects around the Old Town, from the shared sewage system to the restoration of neighborhoods and historic buildings, from the design of pedestrian itineraries and the upgrading of public spaces to the production of documentation, surveys, and data (Abu-Orf 2005: 46). Therefore, the NMP represents a patrimony of collaborative planning between the two sides upon which to build. The first phase of the NMP (1981–1984) dealt primarily with physical improvements, such as infrastructure or building restoration (Ewers 2018: 55). While necessary for the preservation of the

[1] First established in 1979, the Nicosia Master Plan (NMP) represents an inter-communal planning endeavor, aiming at assessing the city's structure and at developing a long-term plan for its coordinated development—in the scenario of a potential reunification as well as in the prolongation of the status quo.

[2] Based at the Home for Cooperation and funded also by the EU and UNDP, the Technical Committee for Cultural Heritage (TCCH) is composed of archaeologists, architects, art historians and town planners from both communities, working on the preservation of Cyprus' multicultural built heritage. See also https://tcchcyprus.com/ and https://www.undp.org/cyprus/projects/support-monuments-great-importance-communities-cyprus-phase-7.

historic fabric and the functioning of the city, the process "did not significantly involve citizenship and was not accompanied by an attempt to overcome mistrust and prejudice" (Casaglia 2020: 177). In other words, the first phase of the NMP focused mainly on the material aspects of city-making from a programmatic point of view (Ewers 2018: 55), with less focus on the social ties and narratives constructed around urban space. On the contrary, the second phase (1985) elaborated on urban strategies for strengthening the city center on a social, commercial and cultural level (see introduction to this volume), which were further materially implemented during the third phase (1986—ongoing). Projects[3] carried out within the NMP framework, though coordinated across the border (Fig. 5), were mostly carried out separately on two sides, both due to restrictions on crossing the Buffer Zone in the 1980 and 1990s, as well as because of different funding coming from different administrations (Ewers 2018: 97). Furthermore, despite being rightfully celebrated as a successful instance of inter-communal cooperation, in recent years the NMP cross-border collaboration has reached a point of stalling (Papallas 2016).

In parallel, civil society actors on both sides are extremely active, and in many instances, have been pressuring governments to achieve a political solution to the "Cyprus issue". Over the past decades, a variety of projects, actions and events has been developed, with various languages and formats, directly or indirectly tackling the division in urban space: the Buffer Fringe performing arts festival, organized yearly by the Home for Cooperation, brings together artists and audiences to "bridge the divide" and discuss themes such as displacement, otherness, and the in-between space (BFPAF 2021); the OccupyBufferZone movement, emerging in 2011 in the wake of the financial crisis, brought together activists from both sides occupying the Ledra-Lokmacı checkpoint, calling for a peaceful solution to the division and denouncing the status quo (Antonsich 2013: 170); the urban environmental Ecopolis festival, carried out in front of the Old Municipal market in 2019, discussed the issue of gentrification and urban commodification in the Old Town in the RoC, calling for "a broader participation of the city's inhabitants in the decision-making processes" (Ioannidou 2019): the Pame Kaimakli festival, organised by the Urban Gorillas collective in the border-zone neighborhood of Kaimakli, focuses on community-based activities including residents on both sides of the border (Urban Gorillas 2022). And it should not be overlooked that the opening of the crossing points in 2003 was a consequence of a popular protest movement in the TRNC.

An existing inter-communal planning framework, then, is already in place, and civil society actors and community-based associations are actively working towards joint development in Nicosia's urban space. However, considering the growing

[3] Priority projects of the NMP within the Walled City included the Chrysaliniotissa (RoC) and the Arabahmet (TRNC) Neighborhood Conservation Projects (RoC); the Selimiyie Improvement Project (TRNC); City Walls, Bastions and Moat (TRNC, RoC, United Nations Buffer Zone); the Mula Bastion Open-air Theatre (TRNC); the Famagusta Gate Open-air Theatre (RoC); the Ledra and Onasagorou Street Project (RoC); and the Kyrenia Avenue and Saray Square Project (TRNC). Other investment projects include Eleftheria Square (RoC); the Pedieos River Landscaping (TRNC); the Tripoli Bastion Parking Garage (RoC); and the Survey of Buildings within the Buffer Zone (Ewers 2018: 114–115).

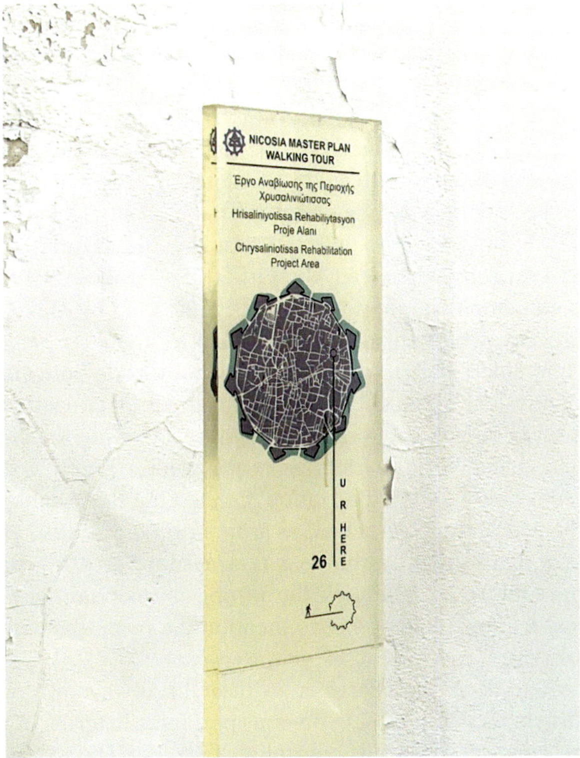

Fig. 5 NMP walking tour map: the only map we came across displaying both sides of the Old Town. Photo by Dafni Riga, 2021

energy for inter-communal action coming "from below", it appears that there is room for scaling up coordination between professional organizations, official inter-communal stakeholders, and civil society actors. There is thus an opportunity for more structural cooperation between the groups of actors: from complementary perspectives, the proposed projects thus tap into the Nicosia Master Plan, with the aim to link official planning with civil society.

2.1.2 Reactivating Under-Used Spaces

The creation of the Green Line impacted Nicosia considerably. The Buffer Zone comprises about 10% of the area of the Old Town (Casaglia 2020: 71), including dozens of structures and most of Ermou street, once the city's main thoroughfare and commercial axis (Papadakis 2006: 99), now a no-go zone. Because of its character as a militarized international border, as well as of its associated memories of violence and conflict, the establishment of the Green Line triggered a process of decay and abandonment in its proximities, affecting the walled city's economy, infrastructure, built heritage and social fabric. Several streets became dead ends, shops and homes were abandoned, and many buildings fell into disrepair (Casaglia 2020: 68, 149). In a sort of vicious cycle, the flight of residents stifled economic activities and prevented buildings maintenance, and the unkempt buildings and deserted streets, in turn, kept newcomers away. Indeed, during our field visits, we witnessed an abundance of residual spaces, from empty plots to abandoned and derelict buildings; not to mention the dead ends produced by the trace of the Green Line, some of which have been re-furnished and utilized by residents as semi-public extensions of their dwellings.

Following a decades-long debate, it is only recently that this process of deterioration has begun to be reversed, thanks to the efforts of both Municipalities to restore and revitalize the area (Abu-Orf 2005: 47; Welz 2017) and with the influx of young populations and migrants, initially attracted by the availability of cheap rents. This process, while raising serious concerns about the ongoing gentrification of the area (Ioannidou 2019), created new centralities on both sides of the walled city in terms of physical urban geography, but also in the collective imagination. The opening of the Ledra-Lokmacı crossing point, reactivating the North–South axis linking Girne gate in the North to Eleftheria square in the South, was a milestone in the process of rediscovery and revitalization, connecting the neighborhoods within the walls with the rest of the city, as well as transforming this dead end into a node of cross-border flows and trade. In the near future, more crossing points could be opened within the Old Town, although the timing is uncertain and dependent on volatile political circumstances—not to mention the recent pandemic-related restrictions on the crossing.

The decaying character of the Old Town, with residual spaces and crumbling structures, while posing obvious problems in terms of heritage conservation can also be considered as an opportunity (Fig. 6). The availability of void, idle or under-used open spaces with the potential for new (more or less temporary) functions, is a significant resource that the proposed projects aim to harness. The prospect of new crossing points opening, and the consequent redefinition of flows and itineraries within the Old Town and around the Venetian Walls means that buildings and spaces in their vicinities have the potential to be re-activated. In this light, the projects presented in the volume build on this promising perspective and recommend actions to counter this process of abandonment and stalling, attracting new inhabitants, functions and users.

2.1.3 Building upon a Shared Socio-Ecological System

Divergent narratives exist concerning the sense of community and ethnonational belonging in Cyprus, both at present and in the pre-partition era. The two sides promote the view of Greek and Turkish "nations" as distinct communities, and some sources claim that "a shared Cypriot identity never developed" (Casaglia 2020: 133). However, there are also sources that highlight the island's long history of multicultural living: for instance, before the rise of nationalism, "Turkish and Greek-Cypriots did not necessarily conceive of each other as distinct communities in ethnic or national terms" (Navaro-Yashin 2012). Contemporary social actors on both sides challenge the essentialist narratives, confrontational attitudes and official interpretations of history

Fig. 6 Interior of an abandoned house, RoC. Photo by Dafni Riga, 2021

mobilized by competing nationalisms (Papadakis 2006: 12; Papallas 2016; Casaglia 2020: 174). Without delving into this debate, what interests us is that Cyprus' socio-ecological system, material culture, and the daily practices of its inhabitants, still hold broad evidence of a hybrid tradition and multicultural identity (Fig. 7). The built and natural environment of the island, and its capital Nicosia, challenges the existing political divide and competing narratives of exclusive heritage, pointing to "the rich living patterns and complex interactions between different Mediterranean communities and cultures over successive generations" (Pulhan and Numan 2006).

This is not to deny cultural heterogeneity within the island; quite on the contrary, we aim to highlight that such existing diversity is not articulated along rigid ethno-cultural lines. In other words, material and immaterial mundane elements—from agricultural traditions to popular beliefs, from typical food to local crafts—are not only shared between the two purportedly antithetical "communities" but refer to a much more complex and multi-layered identity. Looking more specifically at the architectural heritage, traditional construction techniques, building materials, and vernacular typologies are common across the two sides, while the coexistence of elements from various architectural traditions points to a cultural syncretism crafted over the centuries (Pulhan and Numan 2006).

Fig. 7 Men sitting in a coffee shop in the RoC and in the TRNC. Photo by Dafni Riga, 2021

From a socio-environmental perspective, in spite of the man-made border, Cyprus constitutes one single ecosystem (Sadri 2021) (Fig. 8). The island and both "states" rely on the same natural resources and are equally impacted by climatic events (including, of course, climate change) (Hadjinicolaou et al. 2011: 442).

Working on the ecological system, socio-economic interlinkages, and material culture of everyday life in Nicosia is, therefore a way to focus on the shared tangible and intangible heritage of the city and the mixed identities of its people, eschewing divisive and exclusionary narratives. Our strategy thus aims at addressing the conflictive issue of partition and division indirectly, working on "down-to-earth" dimensions of local development. At the same time, having witnessed both heterogeneity and commonality in the material culture of Nicosia's Old Town, we intend to tackle subjects of mixed identities and suggest ways to recover the socio-economic interlinkages between the two sides.

Fig. 8 View of Nicosia with Ledra Street (RoC) and the Turkish flag on the Five Fingers Mountain (TRNC). Photo by Dafni Riga, 2021

2.2 Stapling Strategies

What we define as a stapling strategy is a process that builds upon, and develops, existing socio-spatial interlinkages between the two sides of Nicosia. The process aims at "stapling" or "stitching" together spaces that used to be contiguous and continuous but are now severed by the Green Line. The strategy deals with three complementary macro-themes, namely: (1) the Venetian Walls as a unifying green system; (2) the water system as a socio-cultural infrastructure; and (3) food and trade as a shared patrimony and livelihood. The themes are interpreted and developed with a specific focus on public space and civic use, physically and socially articulating the dialogue between the two sides of the city. The focus of the interventions is, at the same time, on both the physical space and the social relations that evolve around it.

The proposed interventions, therefore, do not constitute a prescriptive masterplan, nor a "all-encompassing" vision for the walled city. Rather, they compose an open framework for strategic, concerted action; they are complementary but not necessarily interdependent. Based on the principles of community-based action planning (Hamdi and Goethert 1997), the projects presented in this book have been developed following an approach that is:

Focusing on starting points, rather than on end states: Building upon the existing patrimony of inter-communal cooperation, the projects aim at coordinating small-scale, short-term interventions conceived as steps towards a joint development of the city, without delineating a predetermined, large-scale outcome.

Incremental, rather than comprehensive, and flexible enough to correct course along its way: Considering the volatile and hardly predictable political circumstances, the interventions focus on the cumulative effects of interventions carried out through a "trial-and-test" method. They are not strictly prescriptive, which makes them adaptable to the changing situation in terms of scale, timing, outcome – accommodating, as the NMP itself, both an open-ended prolonging of the status quo and a possible future reunification.

Based on achievable actions (even if less than best) driven by existing opportunities, hopes and aspirations: the interventions build upon an existing ambitious vision for Nicosia, however, they recognize the limitations and obstacles in terms of budget, capabilities, and political will. They are thus focused on short-term localized interventions, which, however, if coordinated, could scale up in time.

Reliant on local knowledge, skills and traditional wisdom: The main resources that the interventions aim to harness lie in Cyprus' society and its people. We do not intend to introduce ready-made solutions to Nicosia, but propose a framework for enabling local energy, creativity and capabilities to be channelled towards sustainable and shared urban development.

Aiming to encourage tangible outcomes: The interventions are conceived in such a way so as to promote the creation of concrete and visible, if small, changes in the cityscape of Nicosia. This way, the socio-political outcomes of the process will be made visible and appreciable in the everyday life of people.

2.2.1 Guidelines for Interventions Tackling Territorial Fragility

In order to set the framework for implementing "stapling strategies" in Nicosia's Old Town, we developed seven guidelines for three different design proposals to tackle Nicosia's territorial fragilities. These recommendations could be employed to orientate and develop intervention scenarios, ensuring they match the vision, aim, and ethos of the overall strategy.

In brief, the proposed interventions should be:

Inter-communal. Interventions should draw on the historical and contemporary multicultural heritage of Cyprus, stretching beyond "communitarian" boundaries, not only by involving both "sides", with stakeholders from the TRNC and the RoC, but also being open to other groups, identities and possible self-ascriptions, beyond the constructed Turkish/Greek divide.

Participatory and community based. Projects should tap into the collective wisdom of those that know the community best—its inhabitants. This means involving both neighborhood residents and civil society actors at large in the development of the projects' contents and design. Engaging those with a historical perspective, insights into how the area functions, and an understanding of what is meaningful for locals, will help to create a sense of ownership and ensure the success of public projects.

Locally grounded. This means not only respecting the existing built environment and unique heritage that gives the city its identity, but also plugging into the place-based economic circuits, drawing on traditional knowledge, resources available on-site, and, thus, contributing to local development.

Coordinated across the border. Interventions need to be implemented in a coordinated way across both sides of the Buffer Zone. Each intervention should have a counterpart on the other side, ideally happening synchronously. The correspondence between the two should be considered in the material design as well as in the content development and should be made explicit.

Temporary, but not impermanent. Interventions should leave a lasting trace in the city's physical space. Projects can range from temporary installations to periodic happenings, from small-scale urban upgrading to cultural shows or art- and design-based interventions, yet it is encouraged that they include an enduring material component that persists beyond the event lifespan.

Socio-economically sustainable. Interventions should be sustainable from a socio-economic and environmental perspective. This means, on the one hand, avoiding inducing gentrification and displacement. On the other hand, it entails maintaining a simple and adaptable design, with locally available materials that can be replaced easily with low environmental impact. This way, interventions will allow for future enhancement of the space as funds become available and the community becomes more involved.

Inclusive and accessible. The implemented projects ought to be inclusive and open to all individuals regardless of their nationality, ethnicity, creed, affiliation, education level or age. Interventions should seek to create heterogeneous public spaces, attracting diverse social groups. Therefore, accessibility (cultural, physical, linguistic, economic) should be considered.

2.2.2 Coordination of Community and Actors

We envision the projects to be ideally carried out within the Nicosia Master Plan framework, coordinated by bi-communal organizations (such as the Home for Cooperation, for instance) and supported by both Municipalities in terms of political backing and legal-administrative assistance.

The projects could possibly tap into funding from extra-state entities, such as the European Union[4] or UNDP, which besides having the financial capacity, can be perceived as more neutral, at least to some extent. There is a long list of inter-communal projects, in various fields, funded in Cyprus by such institutions—often jointly. These include support to inter-communal Technical Committees—such as the one on Cultural Heritage, working since 2010 on shared heritage preservation, or the one on Crossings, active since 2015 in planning new border crossings—as well as joint infrastructural projects such as Nicosia's new Water Treatment Plant, which opened in 2013 (UNDP 2018). In parallel, civil society-driven inter-communal cultural festivals and initiatives—such as the Buffer Fringe or the Kuir Fest, to name just a few—also benefit from funding and support from extra-state institutions.

Our strategic approach aims at linking urban development planning with civil society, deepening the social embeddedness of the masterplan while also coordinating and scaling up "bottom-up" efforts towards a concerted solution. Therefore, the projects envision the involvement of organizations, associations and individual citizens on both sides in the design of the projects, in the development of their content, and in the implementation and follow-up. Both in the RoC and TRNC, Cyprus has an active civil society. Our hope is that by collaborating on practical urban interventions, non-governmental organizations, the local administration, neighborhood-based associations and professional institutions will build new links and alliances which can orientate Nicosia's future development towards a political settlement.

In this sense, we talk of "community" on a twofold scale. On the one hand, we refer to the people inhabiting and experiencing the neighborhoods where the interventions are going to be located and the potential future crossing points will be opened (the long-time residents, regular users, and temporary dwellers who have a stake in local urban development). On the other hand, we address the broader social collectivity of the city and country (or countries), the heterogeneous polity which engages in the debate over the future of Nicosia and the political division (Fig. 9).

[4] The EU's Aid Programme for the Turkish Cypriot community "aims at facilitating the reunification of Cyprus by encouraging the economic development of the Turkish Cypriot community, with particular emphasis on the economic integration of the island, on improving contacts between the two communities" (European Commission, 2018). For further information, see https://commission.europa.eu/funding-tenders/find-funding/eu-funding-programmes/support-turkish-cypriot-community/aid-programme-turkish-cypriot-community_en.

Among the sectors from where participants could be involved, we identified the following:

Inter-communal organizations and NGOs. There are a variety of associations, with various degrees of formality, size, and reach, which work to address the issue of Cyprus' division and campaign for a fair reunification. They do so from different perspectives and focusing on many different topics, including art, cinema, human rights, environment, public space, or progressive politics. Some of them—such as the Home for Cooperation[5]—are widely recognized inter-communal organizations, composed of members from both the TRNC and RoC. Other organizations, while based in one side, deal specifically with overcoming the division, tackling the conflict, achieving a rapprochement, and to this end, they interact with people and associations on the other side. Bringing in non-governmental associations (both inter-communal and not) to work towards reconciliation is fundamental for developing meaningful interventions and scaling up the social impact of the projects.

Professional organizations. There are many professional organizations operating in Nicosia. They are either sectoral associations, more or less linked to state institutions, or departments of the administration. Thus, they are not inter-communal, but often have a counterpart on the other side: it is the case, for instance, of the Cyprus Association of Civil Engineers (SPOLMIK) in the RoC and the Chamber of Civil Engineers (KTMMOB) in the TRNC, or the Chamber of Architects (Mimarlar Odası) in the TRNC and the Cyprus Architects Association in the RoC. All these organizations comprise professionals who could contribute relevant skills and expertise to the projects, and benefit from interacting with stakeholders on the "other side". Bringing professional organizations across the border to work together on projects in their area of expertise will strengthen the culture of dialogue and engagement between the two sides.

Neighborhood-based organizations and institutions. Zooming into the Old Town, we find various locally based organizations, of different nature. They range from Municipality-run social centers to outposts of confessional groups; from community-based associations to cultural centers; from local branches of international organizations to informal meeting points (such as a bar or a teahouse). Because of their physical proximity to the areas of intervention, the variegated local stakeholders could contribute a lot to the development of projects that are contextualized in the neighborhood.

Local authorities. Local authorities also have an important role to play in the coordination, communication, and implementation of the projects. Relevant municipal offices could be involved at various stages of the project. Local neighborhood representatives (such as the *muhtars*, or neighborhood chiefs, in the TRNC) could also become important local actors. Involving local authorities—both as institutions

---

[5] The Home for Cooperation, opened in 2011, is an inter-communal community center located within the Buffer Zone (in the Ledra Palace area). As stated in its website, "Today the Home for Cooperation has become a landmark building in Nicosia, acting as a bridge-builder between separated communities, memories and visions through its physical presence and its peacebuilding programs benefiting from the transformative power of arts and culture." For more information, see https://www.home4cooperation.info/.

Fig. 9 Fictional Call for Proposals: addressing the local community and actors. *Source* Elaboration by the authors

and through individuals working for them—is instrumental in ensuring the smooth execution of the projects. At the same time, bringing governmental actors into these inter-communal urban interventions is an important step to amplify the impact of the process.

Academia. Cyprus is an important center for higher education (both in the RoC and in the TRNC), and there are various academic institutions, such as the University of Cyprus, the University of Nicosia, or the American University of Cyprus, that offer courses on relevant topics—such as urban design, sociology, heritage conservation, architecture. In many such institutions, there are individuals that carry out research and practical work on the city and the political division. As a site of knowledge production and critical thinking, universities constitute an important source of capabilities and influence: they are a relevant stakeholder to work with. Involving university students will bring young citizens into the projects and contribute to shaping their education and future approach, as professionals and as citizens.

3 Towards a Common Ground

Under the overarching theme of "stapling strategies", the three proposals for intervention scenarios presented in this book collectively address the challenges and opportunities for activating and expanding the common ground between the two sides of Nicosia's Old Town, as presented in this chapter. The projects' main intention is to bridge inter-communal planning policies by involving civil society and strengthening the sense of local community. By achieving this, the projects propose techniques that benefit from the shared immaterial values present on both sides and aim to build upon a shared socio-ecological system in Nicosia's Old Town. At the same time, and in more pragmatic terms, the projects address material values existent in the urban tissue, by proposing the re-activation of under-used spaces.

The project focusing on the Venetian Walls surrounding the Old Town presents three design scenarios for creating a unified system as a green common space that aims to breach the separation line. Based on inter-communal coordination across the border, the strategy proposes flexible, temporary interventions that can be developed in different phases, involving participatory processes and community-based synergies to be achieved through sustainable and independent governance.

By drawing on the importance of community-oriented awareness initiatives, Civic Water tackles the issue of water heritage in Nicosia's Old Town as an opportunity for bridging inter-communal material values and addressing socio-economical sustainability issues. Building on a participatory and community-based approach, this project aims to coordinate existing community and actors in order to develop inclusive and accessible awareness strategies under the theme of water.

Finally, the project Tackling Residuality through Nicosia's Market Heritage aims to bring the city together by investigating market spaces and buildings within the Old Town and proposes scenarios on their reactivation as a means for reconnecting, both materially and immaterially, the two sides. The scenarios are locally grounded, inclusive and accessible, and include a set of policy measures that seek to enhance socio-economical sustainability and inter-communal coordination.

The three projects contribute to the concept of "stapling" by developing strategies that aim to respond to the main criticalities that we have identified in the Old Town of Nicosia (Fig. 10). Each of the projects addresses a number of the guidelines for intervention presented in this chapter, while they all consider coordinating community and actors, by favoring participatory processes, inclusivity, accessibility and community awareness. Finally, the projects are characterized by a high level of flexibility and adaptability: presented in phases, they rigorously tackle temporariness, without however excluding the possibility of an accessible Buffer Zone and the opening of new checkpoints in the future.

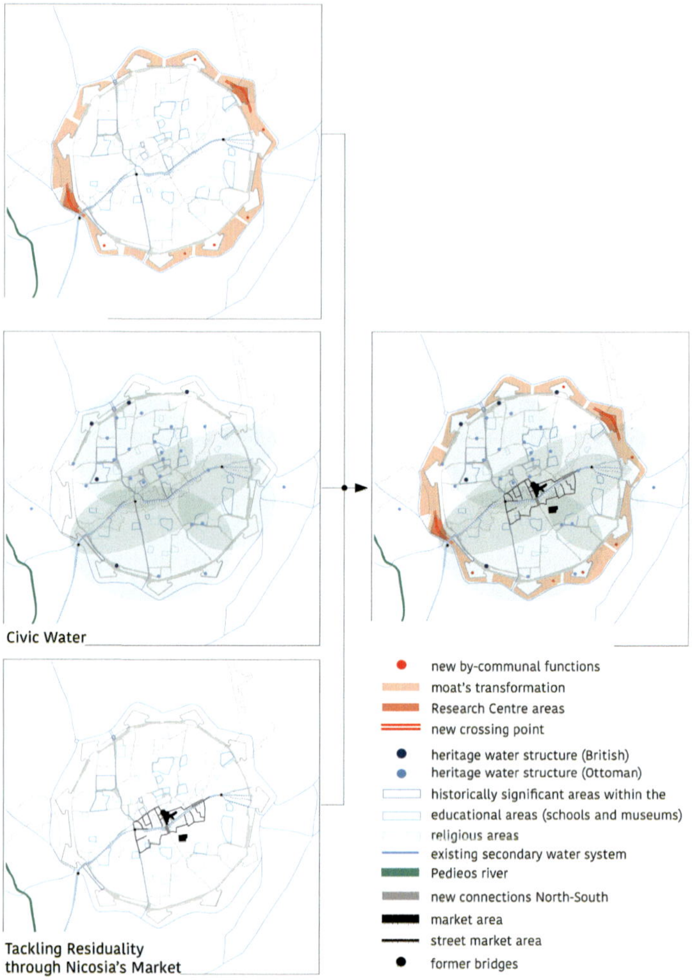

Fig. 10 Map of contemporary Nicosia with the three strategies superimposed, along with respective points of interest. Elaboration by the authors, re-edited with the support of Neofytos Christou, 2023

References

Abu-Orf H (2005) Collaborative planning in practice: the Nicosia master plan. Planning Practice & Research 20(1):41–58. https://doi.org/10.1080/02697450500261707

Antonsich M (2013) 'OccupyBufferZone': practices of borderline resistance in a space of exception. Area 45(2):170–178

Boyle D, Harris H (2009) The challenge of co-production: how equal partnerships between professionals and the public are crucial to improving public services. Available via NESTA. https://beckymalby.wordpress.com/2015/11/08/17/. Accessed 3 Jun 2022

Brambilla C (2015) Exploring the critical potential of the borderscapes concept. Geopolitics, 20(1):14–34. https://doi.org/10.1080/14650045.2014.884561

Buffer fringe performing arts festival (2021) Buffer fringe (2021) Available via home for cooperation. https://www.home4cooperation.info/buffer-fringe-2021/
Buoli A (2020) Borderscap-es/-ing: Reading the Moroccan-Spanish Borderlands Constellations. In: Gaeta L, Buoli A (eds) Transdisciplinary Views on Boundaries. Towards a New Lexicon. Fondazione Giangiacomo Feltrinelli, Milano, p. 21–45
Casaglia A (2020) Nicosia beyond partition: Complex geographies of the divided city (Prima edizione). Edizioni Unicopli
De Certeau M (1984) The Practice of Everyday Life. University of California Press, Berkeley
Deleuze G, Guattari F (1988) A thousand plateaus. University of Minnesota Press, Minneapolis
Dovey K (1999) Framing places. Mediating power in built form. Routledge, London
European Commission (2018) Aid Programme for the Turkish Cypriot community. Available via European Commission. https://commission.europa.eu/funding-tenders/find-funding/eu-funding-programmes/support-turkish-cypriot-community/aid-programme-turkish-cypriot-community_en. Accessed 29 December 2022
Ewers CM (2018) The Nicosia masterplan: historic preservation as urban regeneration. Dissertation, University of Oregon. Available via academia. https://www.academia.edu/37767222/The_Nicosia_Master_Plan_Historic_Preservation_as_Urban_Regeneration. Accessed 3 Jun 2022
Gaeta L (2021) Borders studies: a transdisciplinary and critical enterprise. PowerPoint lecture presented at Politecnico di Milano, Italy. 18 May 2021
Hadjinicolaou P et al (2011) Mid-21st century climate and weather extremes in Cyprus as projected by six regional climate models. Reg Environ Change 11:441–457. https://doi.org/10.1007/s10113-010-0153-1
Hamdi N, Goethert R (1997) Action planning for cities: A guide to community practice. John Wiley
Van Houtum H, van Naerssen T (2002) Bordering, Ordering and Othering. Tijdschrift voor Economische en Sociale Geografie 93(2):125–136. https://doi.org/10.1111/1467-9663.00189
Van Houtum H, Eker M (2015). BorderScapes: redesigning the borderland. Territorio 72:101–107. https://doi.org/10.3280/TR2015-072016
Ioannidou L (2019) Who owns the public spaces? Available via Cyprus Mail. https://cyprus-mail.com/2019/07/07/who-owns-the-public-spaces/. Accessed 3 Jun 2022
Ker-Lindsay J (2019) The Cyprus Problem in an era of uncertainty: establishing a culture of engagement. PRIO Cyprus Center Report 5. PRIO, Cyprus
Lefebvre H (1991) The production of space. Blackwell, Oxford
Mezzadra S, Neilson B (2012) Between inclusion and exclusion: on the topology of global space and borders. Theory, Culture & Society 29(4/5):58–75. https://doi.org/10.1177/0263276412443569
Nail T (2016) Theory of the border. Oxford University Press, New York
Navaro-Yashin Y (2012) The make-believe space: affective geography in a postwar polity. Duke University Press, Durham
Newman D (2006) The lines that continue to separate us: borders in our 'borderless' world. Progress in Human Geography 30(2):1–19
Papadakis Y (2006) Nicosia after 1960: A River, A Bridge and a Dead Zone. Global Media Journal: Mediterranean Edition 1:1–16
Papallas A (2016) In conversation with the Nicosia master plan. Terra Nullius: the logbook. Available via cargocollective: http://cargocollective.com/cdrspapallas/In-Conversation-with-the-Nicosia-Master-Plan. Accessed 3 Jun 2022
Pulhan H, Numan I (2006) The traditional urban house in Cyprus as material expression of cultural transformation. Journal of Design History 19(2):105–119
Sadri H (2021) De-planning Cyprus: A Socio-Ecological Vision. Online lecture, Dialogues in Development, Development Planning Unit, London. 29 April 2021. Available at https://www.youtube.com/watch?v=vqwxVfm_BvE&ab_channel=TheBartlettDevelopmentPlanningUnit
Soja E (2005) Borders, unbound: globalization, regionalism and the postmetropolitan tradition. In: Houtum van H, Kramsch OT, Zierhofer W (eds) B/ordering Space. Ashgate, London, p 33–46
Storey D (2012) Territories: the claiming of space. Routledge, Abingdon

Strüver A (2018) Europeanization in Cypriot borderscapes: experiencing the Green Line in everyday life. Geopolitics 25(3):609–632. https://doi.org/10.1080/14650045.2018.1550390
Sudjic, D (2005) The Edifice Complex, The Penguin Press, New York.
UNDP (2018): Cyprus Projects. Available at https://www.undp.org/cyprus/projects. Accessed 29 Dec 2022
Urban Gorillas (2022): Pame Kaimakli Festival 2022. Available via Urban Gorillas. https://urbangorillas.org/urban-interventions/pame-kaimakli-festival-2022-urban-playground-2/ Accessed 29 Dec 2022
Welz G (2017) European products: making and unmaking heritage in Cyprus. Berghahn Books, New York-Oxford

Tackling Residuality through Nicosia's Market Heritage

Verdiana Peron, Roberta Pellicanò, Aubrey Toldi, Constanze Wolfgring, and Shifu Zhang

Abstract Markets were central in the old city of Nicosia for decades, playing a crucial role in bolstering relations between different communities. Locals developed memories and attachments to these everyday spaces, demonstrating how they are part of a common tangible and intangible legacy, which still serves as a strong point of reference of what Nicosia used to be before the partition. The establishment of the buffer zone in 1964 disrupted the marketspace, transforming once bustling market streets, that were a common ground to the ethnically divided neighbourhoods, into disconnected residual spaces. When considering how to mitigate the existing barriers between Nicosians on both sides, these spaces could play a crucial role. Based on the hypothesis that Nicosia's history as a market city embodies a common, predominantly positively connotated heritage, this study explores its potential in tackling residuality within the area of the historical marketspace and in promoting opportunities for dialogue and encounter. The research was developed through bibliographic research, interviews, and on-site surveys. We propose short and long-term strategies concerning residual spaces and buildings to reactivate and re-center the former marketspace, arguing that these once vivid spaces in the very heart of the city can again assume bridging functions between communities.

V. Peron (✉) · R. Pellicanò · A. Toldi · C. Wolfgring · S. Zhang
Department of Architecture and Urban Studies, Politecnico Di Milano, Milan, Italy
e-mail: verdiana.peron@polimi.it

R. Pellicanò
e-mail: roberta.pellicano@polimi.it

A. Toldi
e-mail: aubrey.toldi@polimi.it

C. Wolfgring
e-mail: constanze.wolfgring@polimi.it

S. Zhang
e-mail: shifu.zhang@polimi.it

© The Author(s), under exclusive license to Springer Nature Switzerland AG 2023

A. Buoli and O. C. Ţiganea (eds.), *Territorial Fragilities in Cyprus*, Research for Development, https://doi.org/10.1007/978-3-031-36076-3_9

1 Introduction

Markets were central—both in terms of spatiality and identity—in the old city of Nicosia for centuries, playing a crucial role in bolstering public communication and relations between and among different communities. As highlighted by the work of Bakshi (2017), the area of and surrounding Bandabulya market, located today in the north between Ermou and Uray street, has been perceived by many Cypriots as the heart of the city, the central location which brought together different communities and fostered the establishment of spatial and social identities. Whereas most other neighbourhoods were divided along ethnic lines, Ermou marketspace, as well as other commercial streets, were places shared by all communities. Starting in the medieval period, the diversity of customers and businesses turned Nicosia's historical centre into a bustling marketspace as activities were not limited to the market building itself but stretched out on both sides of Ermou street and into its backroads, encompassing local artisans such as carpenters and welders. However, in 1964, the establishment of the buffer zone had a tremendous impact on Nicosia's communities and the entire urban fabric. It disrupted the marketspace, dispersing economic activities, destroying existing centralities, and transforming sites of exchange and encounter into residual spaces. The historical centre thus declined as businesses and residents moved out, leaving residual spaces within this once vibrant area.

Yet, as Bakshi illustrates through a collective mapping project published in 2017 (Fig. 1), locals maintain strong memories and attachments to these everyday spaces, demonstrating how Nicosia's marketspaces were much more than merely sites of economic exchanges and competitive activities. They have provided the scenery of a rich collection of both shared and contradictory memories of local histories, of collaboration and conflict. Today, they are part of a common tangible (through built space) and intangible legacy (through products, food, traditions), which still serves as a strong, and to some degree idealized or biased, reference of what Nicosia used to be before the partition. In this sense, they can be understood as a depository of memories and heritage. Our research focuses on Nicosia's old city as a lived place and creates a proposal that builds upon the city's rich market heritage, an *everyday heritage*, hoping to inspire new imaginaries rooted in predominantly positive memories on both sides of the buffer zone. The study seeks to consider more than the singularity embodied through the Green Line and embrace the everydayness of the city, approaching Nicosia as a place where people live, work, and play despite the partition.

We aim to understand the spatial and functional evolution of Nicosia's central marketspace, to recognize the impacts of the city's partition on these spaces, and to identify potential opportunities that the now *leftover spaces* in the very center of the city may provide for Nicosian communities on both sides of the Green Line. While different functions of markets have been well researched in stable political contexts (see, for instance, Watson and Studdert 2006; Morales 2009; Watson 2009; Janssens and Sezer 2013), their roles in conflict-ridden cities have received far less scientific attention. Given the residualization of these spaces in Nicosia, our project builds upon

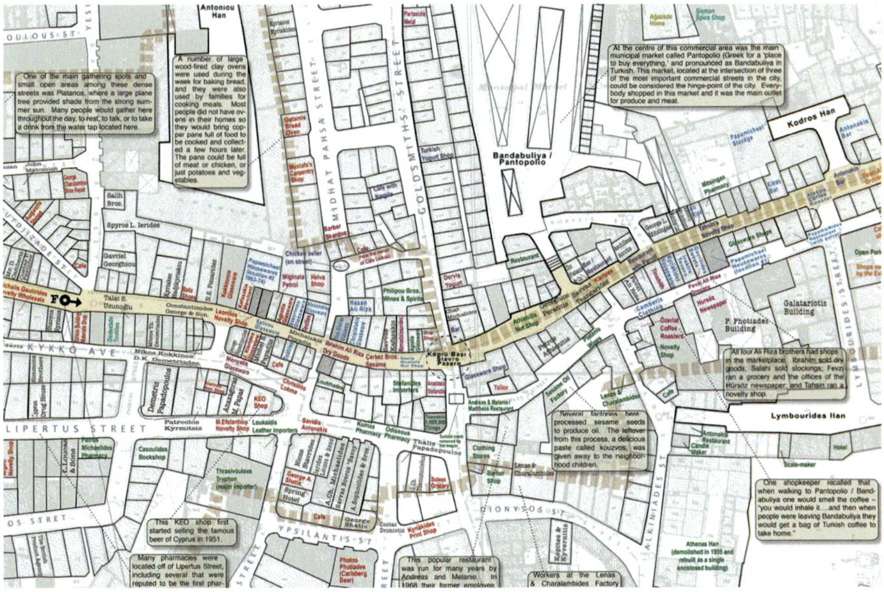

Fig. 1 Nicosia's marketspace before the partition. Reproduced from Bakshi (2017)

the existing theoretical discourse on residuality and extends its conceptualization, applying it to conflictual areas—as that of Nicosia—through the investigation of the former marketspaces lying on both sides of the buffer zone. Differing from presupposed ideas of residual spaces as unused or leftover spaces, our fieldwork in Nicosia[1] demonstrates that there are dynamic activities and communities already present within some of these spaces, showing how they can be sites of vividness and potential. This paper thus proposes occasions for further activating residual spaces near the Green Line and for envisioning ways in which it can become more permeable, as we recognize residuality as a starting point for (re)activation, (re)appropriation, and (re)connection.

The main goals of our fieldwork were to identify dead-end streets and underused or unused spaces along the edges north and south of the Green Line and to further explore their character, types of usage, and their actual *residuality*. Over the course of a week, we conducted a series of surveys at different times of the day, documenting our observations through maps, photographs, and field diaries. Our understanding of the spaces and buildings was expanded through informal interviews. Based on the hypothesis that Nicosia's history as a market city embodies a common and predominantly positively connotated heritage which is shared by different communities living in the city, the paper explores the potential of this heritage in tackling residuality within the area of the historic marketspace and in promoting opportunities for dialogue and

[1] See the introductory chapter.

encounter. Dealing with the issue of residuality, considering its tangible and intangible components, we aim to unpack the following research questions in more depth: How did the Green Line affect Nicosia as a market city? In what ways can markets act as depositories of memory and heritage? How can markets support connections between communities? How can residual spaces be activated and interconnected?

2 Nicosia as a Market City

2.1 *Markets as Spaces of Exchange, Interaction, Inclusion, and Heritage*

Cities and markets are inextricably linked to one another. As cities provide location, demand, and social context to marketspaces, marketspaces offer products, profit, and cultural liveliness to cities (Bestor 2001). Within urban societies, marketspaces can assume various roles, serving as spaces of exchange, interaction, inclusion, and heritage.

Markets are fundamentally sites of economic exchange, serving as locations for commerce, trade, and consumption. While marketspaces are foremost the places where goods are exchanged, these exchanges happen through a process that requires communication and engagement between people: a social act going beyond the mere economic transaction, turning markets into spaces of social interaction. This relationship between the economic and the social functions of markets can be considered from different perspectives during the process, for example, the products exchanged, the explicit and implicit rules of exchange, and the spaces where these exchanges happen. Firstly, the trade of products (i.e., tangible goods) is often accompanied by the exchange of intangible goods (information, stories, languages) between sellers and buyers, which may evoke attachment, sympathy or antipathy to places and cultures (Cresswell 2016; Hiebert et al. 2015). Secondly, the exchange of products does not solely depend on explicit rules such as the mutual agreement over a purchase price (which, differently from other sites of commerce, is not necessarily predetermined but can be the result of negotiation), but also on more implicit and interpersonal factors like building relationships of trust and personal bonding between sellers and buyers over time (Offer 1997), fostering longer-term customer loyalty. Thirdly, the spatial properties of markets (including the arrangement of vendors' stalls) can create inviting atmospheres for customers thereby providing occasions for interaction (Pottie-Sherman 2011). As Watson (2009) argues, the degree of sociability of marketspaces—that is, the extent to which they allow for or foster encounter—largely depends on their physical characteristics and locations.

Moreover, markets tend to be more democratic in their accessibility and more capable of absorbing diversity than many other public spaces. While in many other contexts, encounter and social inclusion are created through targeted efforts, at markets they occur in a rather incidental way, as users "mingle with each other and become accustomed to each other's differences in a public space" (Watson and Studdert 2006: 11). This also becomes evident in conflictual areas, where markets can play ambiguous, and at times even seemingly diametrically opposed roles: they can act as relatively neutral spaces in which everyday practices can be carried out irrespective of political divergences, yet they may also be theatres of violence, becoming reminders of conflict (an example for this dual role is Jerusalem's *Mahane Yehuda* market, Feinberg 2010). The inclusionary capacity of markets is also apparent at the microeconomic level, as they provide economic opportunities and spaces for disadvantaged groups: markets often serve as entryways for the economic activities of those that struggle to access the formal labour market. Entry-level requirements for jobs in markets are usually lower, and language or cultural barriers are less relevant (Knierbein et al. 2012; Hann 2015).

Markets can furthermore be perceived as depositories of heritage, both in their materiality (i.e., built structures, spaces and buildings) and regarding the immaterial traces they leave or entail (e.g., food, know-how, traditions). Heritage has been defined as a group of "resources inherited from the past which people identify, independently of ownership, as a reflection and expression of their constantly evolving values, beliefs, knowledge and traditions" (Council of Europe 2005: 2). Apart from its importance in the dynamics of human life on a macro level (for instance, regarding conflict resolution and education for citizenship, see Council of Europe 2009), heritage plays a crucial—and reciprocal—role in the everyday lives of individuals and communities: just as a space turns into heritage by people recognizing it as such, the same people can be shaped through their everyday interactions with that space. As Giombini (2020) stresses, it is people who transform everyday sites into places of human significance since they are part of the social and lived-in dimension. The recognition of an *ordinary heritage* is linked to everyday practices that take place in specific spaces and are considered authentic by people (Podder et al. 2018).

2.2 *The Market City over History: From Centrality to Residuality*

In the medieval period, the marketspace in Nicosia was located alongside the Pedieos River which crossed the city (Leventis 2005). After the diversion of the river by the Venetians (1489–1570), the new streets built on the riverbed formed a commercial corridor, composed of narrow plots running through the city center, whose backbone was Ermou street (Bakshi 2016). During Ottoman rule (1570–1878), the shops and stands along these streets were the main places of commercial life (Michaelides 2012). The Archduke of Austria, Ludwig Salvator, after having lived in the city for half a year in 1873, depicted the interactions between people in the marketspaces, highlighting the diversity of vendors and customers:

> In all these places the most motley crowd in the world is hurrying up and down, especially before noon; peasants in showy dresses, veiled Turkish women, boys with widely opened eyes. [...] bakers, carrying brown bread on wooden trays, pedlars with cakes, fellows offering dainty little bites of meat to the known purchaser. The most varied scene is everywhere before our eyes (Salvator 1983: 55).

Salvator described Nicosia's markets as usually open structures, covered only with mats or linen rugs, and separated by the types of products for sale (e.g., shoes, copperware, cotton, fish) (ibid.).

During the British rule (1878–1960) the main center was located at the intersection of Ermou and Goldsmiths' street, known as *Stavro Pazaro* (lit. *crossmarket*) in Greek and as *Kopru Başı* (lit. *bridgehead*) in Turkish, since it was the former location of one of the main bridges that had spanned Pedieos river (Bakshi 2017). Most streets within the marketspace did not have official names until 1912 but were recognized for a specific trade and consequently identified according to the presence of manufacturers and merchants who had settled in each of them. According to Bakshi (ibid.), whereas most other neighborhoods were divided along ethnic lines, Ermou marketspace (Fig. 2a), as well as other commercial streets, were places shared by all communities.

In the early twentieth century, the streets still reflected the liveliness of the marketspace, but in the mid-1950s, a sharp shift occurred. The opposition of Greek Cypriots to the British colonial government, which had been intensifying since the 1930s, erupted in the 1950s: Ledra street became the scene of violence between British soldiers and EOKA[2] partisans to the extent that it was named *Murder Mile* by the soldiers (Mousset 1971). The situation worsened in the late 1950s, when acts of aggression between Greeks and Turks occurred, moving from Ledra to Ermou street thus causing damage to the spatial and social structure of the marketspace (Bakshi 2017). Specific actions of vandalism, intimidation and pressure exerted by both Greek and Turkish Cypriot separatist groups (EOKA and TNT,[3] respectively) towards merchants and customers targeted at disrupting the commercial life of Nicosia (Markides 1998; Bakshi 2017). Consequently, the usual commercial ties between Greek and Turkish Cypriots gradually crumbled and sentiments of mistrust rose among the communities (Attalides 1977; Patrick and Bater 1976). Everyday commercial activities were more and more pushed out of the formerly shared marketspace, which was increasingly dominated by the presence of (para)military groups and spatial demarcations.

The ongoing process of social and spatial segregation between Greek and Turkish Cypriot communities was reinforced in 1958 with the de facto establishment of two separate municipalities. In 1964, escalated controversies—with Ermou street being again at the center of intercommunal violence—led to the formation of a buffer zone, creating a formal militarized area of division and thus dissolving the little that was

[2] EOKA (*Ethniki Organosis Kyprion Agoniston*, lit. *National Organisation of Cypriot Fighters*) was a Greek Cypriot nationalist paramilitary organization, founded in 1955 with the aim to achieve an end of British rule in Cyprus and unification with Greece.

[3] TMT (*Türk Mukavemet Teşkilatı*, lit. *Turkish Resistance Organisation*) was a Turkish Cypriot paramilitary organization, founded in 1957 with the aim to counter the claims raised by EOKA.

Fig. 2 **a** Ermou street before the partition (Haigaz Mangoian 1930, Mangoian Bros. Ltd. Archive), **b** Phaneromeni Women Market (Felix Yaksis 1958, Press and Information Office of the Republic of Cyprus Archive, 1081–002-FY[46936]), **c** The former market before the construction of Bandabulya (John Lindros n.d., Medelhavsmuseet Archive, JL355), **d** Bandabulya Market's main façade (Constanze Wolfgring 2021), **e** Shop units in the Bandabulya Market (Constanze Wolfgring 2021), **f** The Market of the Old Town Hall during the interventions of renovation (Constanze Wolfgring 2021)

left of the marketspace. Alternatives to formerly shared commercial spaces were found elsewhere and the social life of the city pulled away from the center, both in the north and the south of the old city (Bakshi 2017). The present state of the urban fabric remains a testimony to this history of abandonment:

> The current material reality of the buffer zone represents the loss of a collective way of life. Here an entire web of streets, buildings, occupants, and sounds has been destroyed. While these streets are still in place, the mud-brick buildings that define their edges are eroding away. Signs have rusted, walls have collapsed, and the life that sustained these marketplace streets is gone (ibid.: 108).

The conflict affected Nicosia's well-known and well-frequented marketspaces directly and in various ways. One of the most popular and long-lasting marketspaces, assumed to have arisen already during the Lusignan period (1192–1489) and enduring until the middle of the twentieth century, was the open-air women's market[4] which took place every Friday, though its location changed over time (Theodotou 2016). For centuries, women from different communities got together, producing woven items and selling textiles, fruits, sweets and cosmetics. In 1924, the market had been moved from the area at the northern end of Ledra street to the vicinity of Phaneromeni church, where a permanent market iron pavilion was built in 1928 (Fig. 2b). Due to the struggles, the market had to be closed in 1958 and was later demolished. The main municipal market building, Bandabulya,[5] on the other hand, has maintained its function as a market to this day, however, it lost its role as a space shared by the city's different communities. The present building, constructed in 1932 (Kiessel 2012), replaced the former market existing at the same site (Fig. 2c), composed of an open square surrounded by some built structures.[6] The project by Odesseas Tsagkarides (Nonument 2023) adopted the architectural language of the typical Turkish bazaar typology, characterized by the presence of stone masonries and a timber wood roof and composed of different blocks featuring small shop units (Fig. 2d, e). Located at the intersection of the most important retail streets (Ermou, Goldsmiths', and Arasta), it had been considered the hinge point of the city (Bakshi 2017). The establishment of the Green Line turned this previously central space into a site north of the buffer zone and, therewith rendering it inaccessible to the Greek Cypriot community located in the south. After the physical separation of the city, the southern part not only needed a revitalization of the urban infrastructure following the conflictual events of the previous years (as did the northern part), but now also lacked a municipal market (Theodotou 2016). In 1967, a new municipal market was therefore built in the area previously occupied by the old town hall. The project of the new building was devised by Stavros Economou, one of the most significant figures of modern architecture in Cyprus, who designed an experimental exposed concrete structure, characterized by mushroom-shaped pillars. Due to its architectural features and its cultural, historical, and social significance, the Market of the Old Town Hall was listed for protection in 2011. The market was closed in 2014, when a project for its renovation and conversion into an innovation centre was approved (Fig. 2f). The exterior area was also regenerated after the decision of the municipality, in 2012, to dismiss the open market located in the surroundings of the building and to move it to one of the eleven bastions of the Venetian walls.

It is important to note that Nicosia's marketspaces have changed (and with them their social functions) due to not only the physical barrier of the Green Line but also

---

[4] The market was called *Genaikopazaron* in Greek, lit. *women's bazaar*, and *Cuma pazarı* in Turkish, lit. *Friday bazaar*.

[5] The market was called *Pantopolio* in Greek, lit. *a place to buy everything*, and *Belediye Pazarı* in Turkish.

[6] This can be noted in the Kitchener map of 1882. The cartographer Horatio Herbert Kitchener had been commissioned by the British colonial government of Cyprus to conduct the first trigonometrical survey of the island (Sylvia Ioannou Foundation and Harokopio University 2022).

the co-emerging identity-based, psychological, and administrative barriers. Although Nicosia's marketspaces seem to follow similar historical accounts, these accounts began to diverge once Ermou street became engulfed in the Green Line. Many of its once-connected streets therefore became dead ends that now carry countering political narratives (Casaglia 2020). Although the youth express more interest in ideas of *togetherness* (A. Grichting, personal communication, 15 Feb 2021[7]), there are still many trust-related issues when it comes to bolstering interisland exchanges (CyprusInno 2020). Though these concerns may differ between sides, they continue to fuel one another; for example, Greek Cypriots deny and hide their clients due to fears of being pillarized by their own communities and Turkish Cypriots fear of being treated as inferior, resulting in feelings of resentment (Hatay et al. 2008). Due to various economic and logistical restraints—such as, but not limited to, political and economic isolation of the north (Katircioğlu 2010), restricted telecommunications (Hatay et al. 2008), or even the lack of knowledge of the Green Line's regulations (CyprusInno 2020)—Cypriots remain legally restricted in further developing economic and social engagements between the two sides. These few examples demonstrate that even if the materiality (e.g., the fencing) of the Green Line could be eventually removed, other barriers would remain.

2.3 Exploring Residual Spaces within Nicosia's Marketspace

> As cities age, they get used and abused by people who live in them. We sometimes patch what we damage, but other times we let things fall apart. As a result, lost urban environments are a hodgepodge of haphazard fixes and odd vestiges. Yet pointless leftovers and accumulated remnants are just as much a part of the city as thoroughly considered, still-functional objects. Such imperfect items are not always the prettiest examples of what we can make, but they perfectly represent our flawed and complex humanity (Mars and Kohlstedt 2020: 33).

The lack of attention towards residual spaces seems to be rooted in their very nature, as they are spaces that are hard to grasp, to demarcate and to classify. For instance, these spaces could be places or buildings that have been abandoned, infrastructures that have become obsolete, the negative/leftover space in between designed or curated spaces, and dead ends. This intangibility is also reflected in the theoretical discourse on these spaces. There is no uniform understanding of what residual spaces are, of the role they assume within the urban fabric, or what potentials they might entail. However, attempts have been undertaken by some scholars, turning to different terminologies and stressing diverse aspects of the phenomenon: leftover spaces, void spaces in the background of everyday life that "may [have] outlive[d] their purpose" (Rickles 2019: 29), *terrains vagues* that are "mentally exterior in the physical interior of the city" (Solà-Morales 1995: 120) or even "paradoxical spaces [...], neither planned nor designed [...], often results of mistakes, coincidences, neglect" (Barron

[7] See the introductory chapter.

and Mariani 2013: 135). Inspired by these fragments, we established a working definition that builds upon four dimensions through which the process of residualization can be understood and addressed:

- a *time* dimension: Residualization occurs over time. It is usually a process shaped by the cumulative effects of regular and irregular developments. However, it may be initiated (or consolidated) through a sudden disruption (such as a natural or man-made disaster, a military conflict, etc.);
- a *mental* dimension: Residualization affects the meanings and images of spaces, leaving emotions, stories, and memories as residues of former connotations attributed to these spaces;
- a *spatial* dimension: Through the process of residualization, spatial qualities and spatial relations change (e.g., spaces are abandoned and deteriorate, formerly central spaces become marginal, former connections become dead-end streets, etc.);
- a *usage* dimension: Through residualization, the use and purpose of a space changes or dissolves.

Considering that residualization is a process, we argue that residual spaces can be "de-residualized" and refilled with purpose and meaning through (re)activation, (re)appropriation and (re)connection. Because of this, there is ample reason to dedicate attention to residual spaces, as they entail a variety of potentials, yet are mostly underestimated and unused. The absence of design and curation gives them a certain authenticity that carefully planned spaces often lack, and an openness for appropriation. Chase et al. (1999) emphasized that the emptiness of residual spaces gives users the freedom to produce and reproduce place meanings through spontaneous appropriations. Moreover, residual spaces provide different opportunities for usage, experimentation, and innovation: underused spaces, caught in the fractures of bustling urban environments, show a more privileged atmosphere for projects (Scoppetta 2010) and can be spaces for informal, self-organized or temporary uses (Willinger 2007).

The residual spaces in Nicosia's historical centre have been shaped not only by regular processes of residualization over time, but also through the forceful conflict, still evident today in the form of the Green Line. Due to the establishment of the buffer zone, streets have been cut off and deprived of their uses and relations with other spaces, creating a broken network of dead-end streets, of void spaces and of buildings vacated during the conflict and not reused ever since. These spaces and buildings allow for an ambiguous interpretation as forgotten landscapes, simultaneously serving as reminders of forceful disruption and as sites of potential. Given the historical and the potential value of Nicosia's residual spaces—in view of a (at this time hypothetical) scenario of reunification, but also considering a scenario in which the status quo remains—it seems promising to address the question of how they can be reactivated and filled with new uses and meanings. If physical barriers might be partially or entirely lifted, today's residual spaces will be at the very city centre and, therefore, valuable spaces for communities to interact.

As we explored the edges of the Green Line within Nicosia's old city, we collected information, impressions and insights, allowing us to identify residual spaces that demonstrate potential for activation or connections between the north and the south. We found that the character of these former marketspaces presents both unique challenges and features. The Southern edges of the Green Line can be perceived as much less inviting due to the noticeable military presence. Many constructions are in a state of decay or already in ruins, although a significant number of buildings have been restored in the recent past (Fig. 3a). These interventions often targeted the facades only, leaving the interior of the buildings untouched, thereby making it difficult to distinguish between used and vacant buildings. While many renovated buildings seem vacant, many of the ones still in need of repair are used by craftsmen. Thus, south of the Green Line, the contrast between *the new* and *the old* is evident, and the transition between them is quite abrupt. For example, at the intersection of Pentadaktylou and Ermou street (Fig. 3b), one can perceive the increasing decay of the buildings inside the buffer zone to the west and, in contrast, of renovated buildings accommodating highly priced stores, cafés, or shopfronts of young creatives on Ermou street to the east.

Along the northern edges of the Green Line, the use of buildings and spaces is more heterogenous, with a mix of functions such as bars, restaurants, fabric stores, workshops and tailors. Residential uses near the buffer zone contribute to a higher density and frequency of users as well as to other types of usages of the dead ends, such as informal semi-private patios (Fig. 3c). This heterogeneity is also reflected in the demographic structure, with an overall higher diversity and a larger share of younger population groups. The variety of uses and users creates a more vivid and lively atmosphere that spills into the dead ends.

Despite these differences, there are several shared challenges and/or features between the two parts of the city. Many buildings have been abandoned and are in advanced material decay but, although this affects both sides, the extent of deterioration (Fig. 3d) is more widespread in the north, where the repair of buildings seems to be a major challenge. However, some interventions have occurred, as in the Bandabulya market, which had been renovated between 2010 and 2012 with EU funds (UNDP 2013). The works included the replacement of the timber roof, the application of a new layer of plaster and the rehabilitation of stalls. Another remarkable similarity concerns the presence of workshops and traditional craftsmanship, which can be considered a valuable resource for livelihood, culture and economic well-being. South of the Green Line, these workshops (Fig. 3e) are managed by a generation soon approaching retirement, and they currently seem to be the main source of liveliness along the edges of the buffer zone. By contrast, in the north (Fig. 3f), a wide range in the age of merchants, as well as in the types of businesses operating in the area, can be observed. Lastly, many residual spaces exist along both sides of the Green Line, though their presence is more evident in the south. This could be due to several reasons, such as the more noticeable military presence and the lack of a residential population. During informal interviews, local stakeholders raised concerns about the loss of once public spaces, such as the former municipal market, and processes of gentrification clearly taking place in the south. Moreover, they stressed the need of attracting regular residential users to the area and establishing facilities for daily needs, e.g., grocery stores, schools, and markets.

Fig. 3 **a** Restored building south of the Green Line (Constanze Wolfgring 2021), **b** The intersection of Pentadaktylou and Ermou street (Verdiana Peron 2021), **c** A semi-private patio north of the Green Line (Constanze Wolfgring 2021), **d** An example of the state of decay north of the Green Line (Constanze Wolfgring 2021), **e** Inside a carpentry workshop in Lidinis street (Constanze Wolfgring 2021), **f** Inside a brewery north of the Green Line (Constanze Wolfgring 2021)

Overall, the area surrounding the buffer zone is characterized by strong contrasts not only between the north and the south, but also within each of its parts. This diversity allows us to imagine different scenarios of potential futures of these spaces, but the presence of workshops, the role of residential uses and the functions of public spaces raise questions about how such futures could accommodate all.

3 Towards De-Residualization: A Project Proposal

3.1 *Project Goals*

The role of Nicosia's marketspaces has changed over time, in part due to the conflict, but also due to regular processes of economic and spatial transformations that are happening globally. These shifts allow for new layers of meaning to be added to Nicosia's residual marketspaces, nurtured by the symbols and stories these spaces still contain (Casaglia 2020). Focusing on residual spaces around and in between the two main markets, Bandabulya in the north and the prior Market of the Old Town Hall in the south (Fig. 4), the goal is to demonstrate the potentials for engagement between different Nicosian communities. As this area is of interest to many stakeholders, our proposal is situated not only in currently existing plans and interventions, but with an eye towards the future, conceived as practical steps forward that are applicable to various future visions of the city. One key aim is to propose occasions for re-activating residual spaces near the Green Line, with an emphasis on addressing how long-standing craftsmen and new target groups can use them in complementary ways. The other main objective is to address the Green Line itself, envisioning how it can become more permeable and how the former marketspaces can regain their lost centrality.

These aims will be addressed through three proposed strategies that tackle both tangible and intangible dimensions of residuality: (a) activating residual buildings; (b) activating residual spaces; and (c) reconnecting Nicosia's former marketspaces. The first two strategies are devised from short-term and long-term perspectives, recognizing that temporary interventions can create opportunities for more established ones. The third strategy is a hypothetical scenario that considers a physical connection between the (then) re-activated spaces on both sides of the Green Line. The common thread guiding all three strategies is Nicosia's rich market heritage, both in a spatial sense (focusing on former market buildings and spaces) as well as in a thematic sense (drawing from the legacy of the past marketspaces).

3.2 *Activating Residual Buildings*

The first strategy tackles residuality in its material dimension through two main measures: attracting people to *come and see* and introducing complementary policy measures. Informal interventions create opportunities for residual buildings to be flexibly re-imagined and consciously re-purposed. Complimentary policy measures provide a framework to enhance these interventions and pave the way to more long-standing material solutions. The approaches applied here are inspired by *what works already*, recognizing much is happening within these spaces, and envisions *what could work*, taking a step forward that builds from ongoing everyday practices.

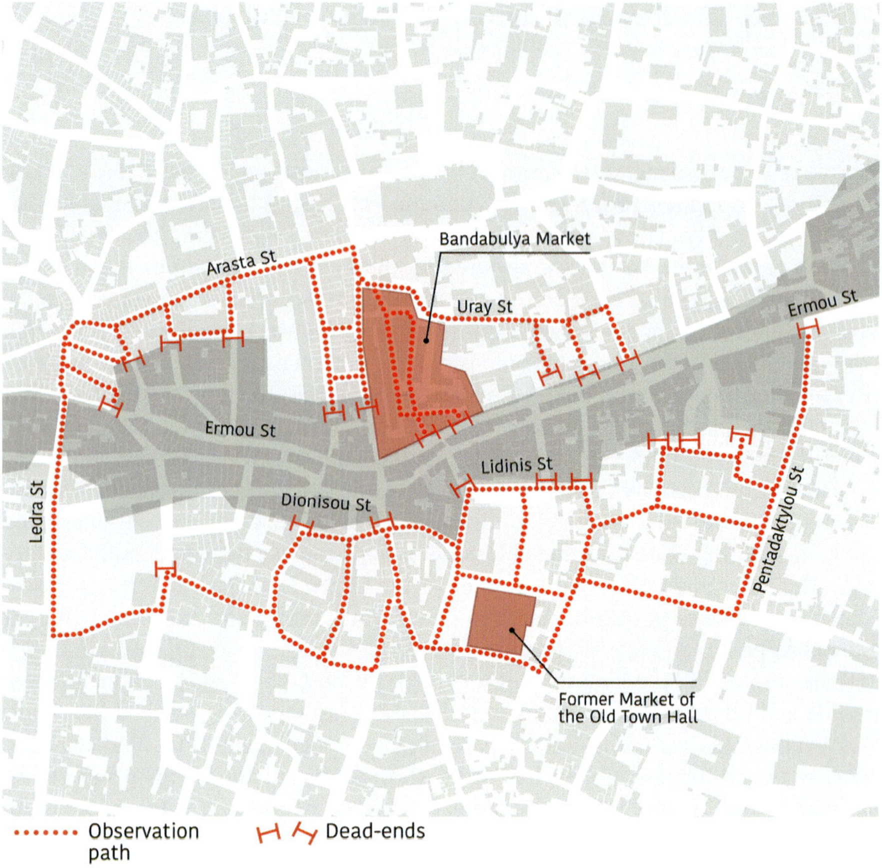

Fig. 4 Map of the focus area. Elaboration by the authors

The first approach is to encourage local craftsmen to exhibit their products and know-hows. During our walking surveys near the Green Line, many artisans were eager to show us their shops, tools and traditional machinery while openly sharing their stories and memories. We believe that these long-existing workshops—directly linked to Nicosia's market heritage—are valuable resources that should be maintained. Most of the craftsmen have been operating their shops for decades, and many are approaching an age of retirement with no legal standing over what would happen to their current workshops. The disappearance of this *living heritage* through retirement or displacement of those that have maintained this marketspace trade—thereby keeping these areas active over the past decades—would be a great loss. From our perspective, these workshops are an opportunity to engage with a meaningful part of Nicosia's market heritage and to re-interpret it for the future. Therefore, we propose to showcase the craftsmen's products and know-hows through open houses, in which interested visitors are invited into their workspaces, or through coordinated events,

Fig. 5 Unused spaces alongside the Green Line (Constanze Wolfgring 2021)

like street markets. These initiatives would not only showcase their crafts at the sites of production but also encourage commercial and personal connections.

Secondly, we propose to introduce events and pop-ups to the area. During our fieldwork in Nicosia, we noticed the presence of active local communities organizing various public events, as advertised through flyers posted throughout the old city. As these activities are currently occurring in more vital and established spaces, our second approach is, therefore, to encourage similar cultural events and pop-up activities in the focus area. Alongside the Green Line, there are several unused spaces and vacant buildings with the potential to host open-air activities, like the large vacant space at the end of Uray street in the north or the underused parking lot on Trikoupi street in the south (Fig. 5). This could not only bring new energy to the neighbourhood, but also direct wider attention to the ongoing activities and the area's existing qualities.

Thirdly, we want to stress the importance of minimizing the side effects of maintaining workshops within the old city centre. There are potential obstacles to activating an area largely composed of workshops, such as safety and noise during the workday. However, we believe that an approach encouraging a temporally stratified use of these spaces can be fruitful, understanding that space can serve different yet complementary uses at different times of the day. The potential success of this suggestion can be seen at the Charátsi bar on Lidinis street. During the daytime, it is a traditional workshop street permeated with a laid-back atmosphere, while on certain nights, young people gather to enjoy drinks and music. This coexistence highlights how *the young and old*, or *the new and existing*, can occupy the same streets without disturbing the rhythm and lifestyle of the other. Keeping in mind that space can serve different purposes throughout a given day, more types of these complementary uses can be identified within the area.

While temporary on-site measures are crucial for creating new spatial imaginaries, an accompanying policy framework is indispensable to sustain their longer-term effectiveness. We, therefore, propose policy measures that tie in with the pre-mentioned measures, aimed at providing incentives for investment around the buffer zone while at the same time preserving existing qualities and resources.

The first proposed policy measure would establish subsidized apprenticeships and intergenerational partnerships between the craftsmen and young trainees. Many workshop owners have fixed, long-standing rental contracts, and have been renting the same shophouses for more than fifty years. As they are reaching retirement age and these contracts cannot be passed on or inherited, the future of these workshops is called into question. The current projects within the area, such as the transformation of the Market of the Old Town Hall, are focused on attracting young creatives. We are unaware of current initiatives underway to preserve the existing workshops. Building from potential connections sparked at the open houses and showcases, subsidized apprenticeships and intergenerational partnerships can go a step further to provide employment opportunities and preserve these long-standing businesses. Subsidized apprenticeships educate future craftsmen, whereas subsidized intergenerational partnerships then aid in maintaining affordable rental contracts for current businesses once the current owner retires.

The second proposed policy measure would promote market investment in the area. Attracting permanent residents to the area is challenging for several reasons. As political uncertainties continue and property-related questions remain unsolved, investment is disincentivized, leaving many buildings in need of repair. Moreover, the area around the Green Line is not perceived as inviting for residential use or daily activities due to the military presence, which is seen as a constant reminder of the conflict. Lastly, housing options in the old city, particularly for middle-class residents, seem to be rather limited. In the past decades, two housing rehabilitation interventions have been carried out as part of a twin project funded by UNHCR and UNDP within the old city in neighborhoods near the buffer zone: Arabahmet in the north and Chrysaliniotissa in the south (UNHCR 1995). The aims were to attract new low to middle-income families, increase the number of housing units and improve the quality of services and public spaces. Continuing the efforts in expanding the housing market and boosting investment, the *1 Euro Houses* model might be a suitable reference for Nicosia. The scheme, promoted by several European municipalities,[8] aims at tackling abandonment and building decay in marginal areas through the sale of houses at low prices in exchange for the commitment to renovate and register a residence. These approaches have demonstrated their potential to re-attract people to depopulated areas (Giuffrida et al. 2020). Other similar models could be applied, such as subsidizing rents (for instance, a rent waiver for an allotted amount of time in exchange for self-renovation). Not only could this bring residents back into the area, but also, it could encourage upgrades in the current decaying infrastructure. This scheme could also be beneficial for the shophouses along the Green Line, allowing the new owner or renter to use the ground floor for business and the first floor for housing. It should be noted that the ownership-related and continuing political uncertainties, especially along the Green Line, are highly contentious issues that could impede on using this measure in some instances.

[8] For instance, in Italy, mostly in inner areas (see Case a 1 Euro 2022), but also in larger cities such as Liverpool, UK (see Liverpool City Council n.d.).

We recognize these measures could have a direct impact on the management of the historical built environment. Thus, when reflecting on attracting residents to this area and reusing the historical buildings, crucial aspects that need to be considered are their architectural qualities, construction features, their state of conservation, and their capacity to be testimonies to the history of the city. The evaluation of their current conditions and of potential reuse interventions, as well as the assessment of their cultural significance, require the contribution of experts. However, new owners/renters could actively support the maintenance by monitoring the general conditions and reporting the detected criticalities to professionals in buildings conservation. The constant use of the built heritage and its continued maintenance is indeed crucial in reducing the extent of needed interventions in the future.

Overall, to better inform these proposed policy measures, we call for an inventory of the buildings within the old city. International organizations have been raising awareness and collecting information on the state of the built environment within the Green Line (e.g., in 2013, Europa Nostra has shortlisted Nicosia's buffer zone as one of the seven most endangered heritage sites in Europe, Europa Nostra 2013, EIB Institute 2013). However, a comprehensive dataset of the historic built environment at the edges of the Green Line does not exist to date. Therefore, it is difficult to assess the features and qualities of the buildings, as well as their current conditions, uses, ownership, etc. Such an inventory effort could be coordinated between and led by both municipalities and carried out with the help of professionals, trained volunteers, and institutional partnerships.

3.3 *Activating Residual Spaces*

The second strategy focuses on the activation of residual spaces. This strategy is envisioned to complement the first, as it promotes the exploration of these spaces beyond time and location-specific events. The main measure we propose is a thematic walking path of the (former) market buildings and their residual spaces, highlighting the social and historical significance of past and existing marketspaces in the old city. In Fig. 6, we identified these marketspaces, providing reference points for a path that can be incorporated into organizations' or institutions' broader strategies. This measure could involve both analogue and digital tools. The area of the walking tour can be explored through information panels, paper maps and booklets, providing visitors with information about these spaces, activities and ongoing initiatives. These materials can be connected to digital platforms, such as an app or website, through a QR code printed onto these resources, adding a digital dimension to one's experience. We imagine that these platforms will serve both locals and tourists in several ways. For instance, the digital platforms' design could allow users to build personalized tour paths around the Green Line and add localized pictures, real-time impressions and comments. The platforms could contain a news section to provide information about events, permitting local organizations and institutions to directly contribute. The goal is to provide physical and digital spaces for all to engage with Nicosia's market heritage in a discursive and interactive way, raising awareness of its cultural and historical significance.

There are several local institutions actively promoting projects aimed at enhancing heritage using both traditional and innovative tools, which could serve as gatekeepers or collaborators for such a measure. Museum Lab, a department within the CYENS Research and Innovation Center, would be a highly competent partner for the implementation of such a project. One central objective of the Museum Lab is to "to explore how technology can help expose different layers of history in places of contested heritage and help visitors/users negotiate difficult or awkward heritage"

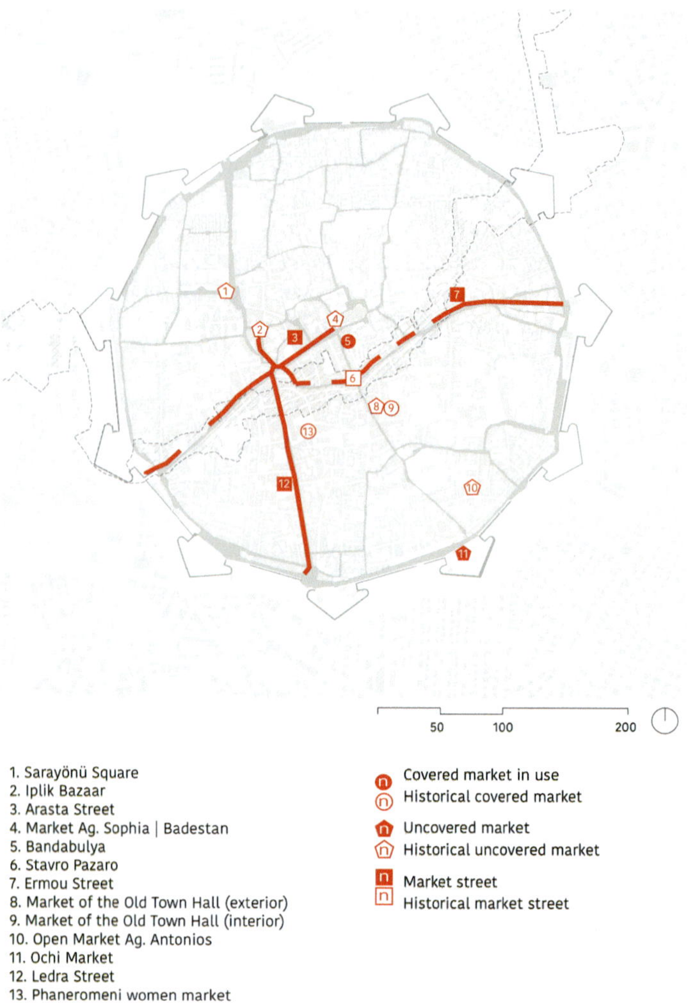

Fig. 6 Map of Nicosia's marketspaces. Elaboration by the authors re-edited with the support of Neofytos Christou, 2023

(Cyens 2021). Another valuable partner would be the Association for Historical Dialogue and Research (AHDR 2021), which produces multi-disciplinary research, policy, and educational materials. Their project *Nicosia. The story of a shared and contested city* consists of an interactive website containing maps of the city's transformation during the late nineteenth and twentieth centuries. As a multi-communal, non-profit, non-governmental organization established in Nicosia in 2003, its mission is to provide access to learning opportunities to individuals, irrespective of personal, religious or cultural backgrounds. The involvement of local actors is crucial for the feasibility of this proposal as they provide concrete support and know-how for its realization.

3.4 Reconnecting Nicosia's Old Marketspaces

Building upon the municipalities' idea to open a new crossing point between the two markets (connecting Trikoupi street in the south with Kuyumcular street in the north), we propose to consider an additional pedestrian crossing point further east, connecting the intersection of Ermou and Pentadaktylou street in the south with Kutuphane street in the north (Fig. 7). This intersection is centred within the sites of our proposal and might therefore contribute to reconnecting Nicosia's old marketspaces on a wider scale. Being the former location of one of the main bridges at the historical centre, which spanned the Pedieos River, this new crossing point could be understood as a new or re-imagined bridge, a reminder of its former function as a connection between the two sides.

Recognizing the very hypothetical nature of this proposal, the implementation of the other interventions might be treated independently from this one. However, in such a scenario, the three pedestrian crossings in the very heart of the city (including the one already existing in Ledra street) would allow for easier accessibility to and from both sides, incentivizing exchange and interactions and re-establishing a centrality that has been lost.

A step towards the realization of such a scenario could be the temporary opening of an enclosed extraterritorial space within the buffer zone at the site of one of the two proposed crossing points (between Trikoupi and Kuyumcular or between Ermou and Kutuphane street), accessible from both sides under the auspices of a bi-communal committee. Such a space could serve as a site for a variety of commercial activities (such as farmers' markets, arts and crafts markets, and flea markets) or events (like concerts, readings, and street food festivals) to promote encounters and exchanges.

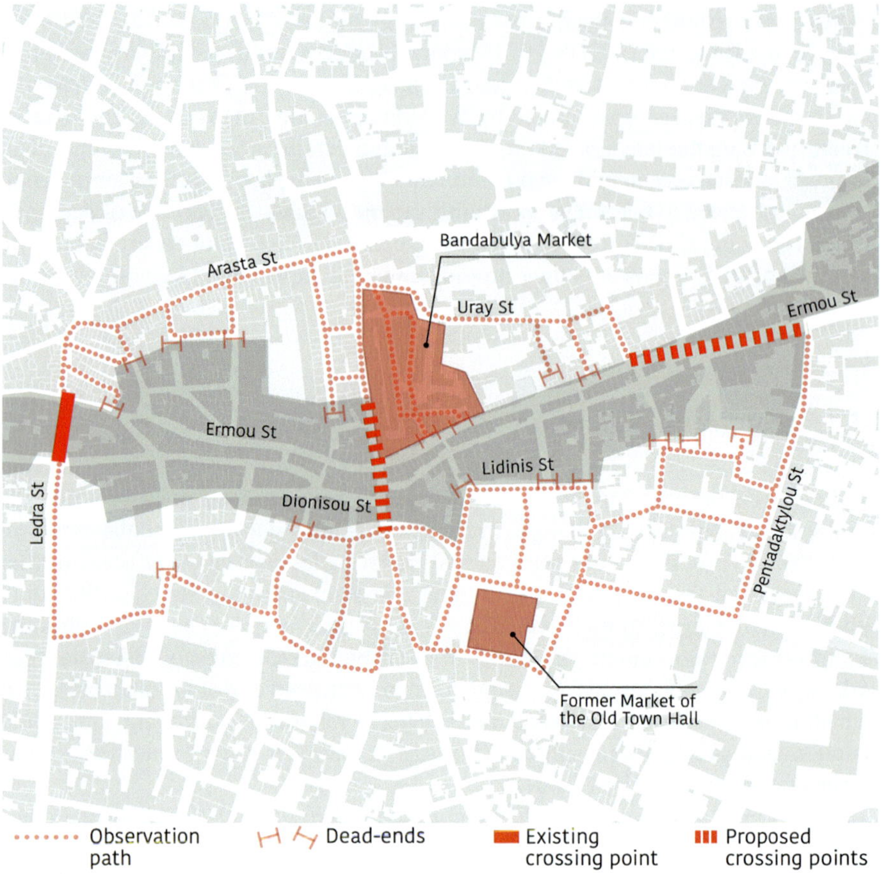

Fig. 7 Proposed crossing points. Elaboration by the authors

4 Concluding Remarks

Our research focuses on Nicosia's old city as a lived place, embracing the everydayness of the city and approaching it as a place where people live, work, and play despite the partition. Picking up on the city's rich market heritage—one that is rooted in predominantly positive memories on both sides of the buffer zone—we hope to create proposals able to inspire new imaginaries for spaces that, once vivid and shared, today seem largely overlooked. Building from the theoretical discourse on residuality, our study extends its conceptualization and applies it to conflictual areas through the investigation of the former marketspaces lying on both sides of the Green Line. The fieldwork revealed the presence of various activities and actors within some of these spaces, demonstrating the limitations of the pre-supposed conceptualization of residual spaces as spaces of abandonment and voidness. We realized that, indeed, processes of reappropriation and reactivation exist at the edges of the buffer zone (albeit on a small scale and in an informal manner), as well as commercial activities that had never left, predating the conflict.

This is, for instance, demonstrated by the presence of highly diversified small-scale craftsmen located within and surrounding the dead ends. The craftsmen occupying these spaces, like welders and carpenters, belong to labour groups commonly pushed to peripheral areas, typically not located within a historic city centre. This unique spatial positioning of small-scale craftsmanship in Nicosia is a valuable resource that many cities try to incentivize. Globally, and also in Nicosia, there is a (re)developmental focus on attracting young, mobile creatives into city centres, which is reflected in urban initiatives, designs, and policies. While this is a meaningful goal, we strongly believe that integrating the existing structures and users into these considerations, interweaving the old and the new, can contribute to the economic and social liveliness of Nicosia's old city and to maintaining its character.

Although we benefited from the freedom of an outsider's perspective, we remain aware of the limitations it entails, even more so in a context as complex as that of Nicosia. We, therefore, would like to emphasize that our proposal is not intended to provide solutions. Rather, we aim at putting practicable ideas on the table that might kick off occasions for encounters and interactions between Nicosians of different backgrounds. As aforementioned, we believe a more thorough knowledge of a site helps interventions be respectful of the vocation of the places, avoiding self-referential projects or proposals taken out of context.

There are many ways in which this research could be further developed. First and foremost, the collection of solid data on vacancies, the state of buildings and ownership structures could put it on stronger grounds and lead to more specific actions. In addition, targeted involvement of the stakeholders in the area (e.g., interviews or focus groups with shop owners, craftsmen, residents, etc.) could provide a better understanding of needs and future perspectives. Another crucial line of research is to explore further schemes with which ordinary residents and complementary infrastructure can be attracted to the area. Affordable housing is a promising (but certainly not the only) starting point. We advocate that such research should be practicable and produce useful and usable insights that might feed concrete projects or initiatives, believing that it is important to *get our hands dirty* to spark positive change.

Acknowledgements This essay is the product of the joint research work and editing of all authors.

References

AHDR (2021) About us. https://www.ahdr.info/about-us. Accessed 18 Oct 2021
Attalides M (1977) Cyprus reviewed. Jus Cypri Association, Nicosia
Bakshi A (2017) Topographies of memories. A new poetics of commemoration. Palgrave Macmillan, Cham

Bakshi A (2016) Trade and exchange in Nicosia's shared realm: Ermou Street in the 1940 and 1950s. In: Bryant R (ed) Post-Ottoman coexistence: Sharing space in the shadow of conflict. Berghahn Books, New York, pp 107–126

Barron P, Mariani M (eds) (2013) Terrain Vague: Interstices at the edge of the pale. Routledge, London

Bestor TC (2001) Markets: anthropological aspects. In: Smelser N, Baltes P (eds) International Encyclopedia of the Social and Behavioral Sciences, 1st edn. Elsevier, Oxford, pp 9227–9231

Casaglia A (2020) Nicosia beyond partition: Complex Geographies of the Divided City. Unicopli, Milan

Chase J, Crawford M, Kaliski J (eds) (1999) Everyday Urbanism. The Monacelli Press, New York

Council of Europe (2009) Heritage and beyond. Strasbourg: Council of Europe Publishing, Strasbourg

Council of Europe (2005) Council of Europe Framework Convention on the Value of Cultural Heritage for Society, Faro, 27 October 2005. Council of Europe Treaty Series n° 199. https://rm.coe.int/1680083746. Accessed 09 Jan 2023

Cresswell T (2016) Hospitality and migrant memory in Maxwell street, Chicago. In: Molz GJ, Gibson S (eds) Mobilizing hospitality: The ethics of social relations in a mobile world. Routledge, Abingdon/New York, pp 47–62

Cyens (2021) Museum Lab. https://www.cyens.org.cy/en-gb/research/pillars-groups/human-factors-design/museum-lab/. Accessed 20 Oct 2021

CyprusInno (2020) About us. https://cyprusinno.com/about-us/. Accessed 17 Nov 2021

EIB Institute (2013) The buffer zone of the historic centre of Nicosia, Cyprus. Technical Report. http://www.europanostra.org/wp-content/uploads/2017/04/7ME-2013-CY-Nicosia-report.pdf. Accessed 30 Dec 2022

Europa Nostra (2013) The buffer zone of the historic centre of Nicosia, Cyprus. https://7mostendangered.eu/sites/the-buffer-zone-of-the-historic-centre-of-nicosia-cyprus/. Accessed 30 Dec 2022

Feinberg M (2010) Machane Yehuda. Master's thesis, University of Waterloo

Giombini L (2020) Everyday heritage and place-making. ESPES 9(2):50–61

Giuffrida S, Trovato MR, Strigari A, Napoli G (2020) "Houses for One Euro" and the territory. Some estimation issues for the "Geographic Debt" reduction. In: International Symposium: New metropolitan perspectives. Springer, Cham, pp 1043–1052

Hann J (2015) Märkte als Katalysatoren für eine sozialraumorientierte Stadtteilentwicklung: eine Untersuchung des sozialen Mehrwerts anhand zweier Fallbeispiele in Wien. Master thesis, TU Wien

Hatay M, Mullen F, Kalimeria J (2008) Intra-island trade in Cyprus: Obstacles, oppositions and psychological barriers. International Peace Research Institute, Oslo

Hiebert D, Rath J, Vertovec S (2015) Urban markets and diversity: towards a research agenda. Ethnic and Racial Studies 38(1):5–21

Janssens F, Sezer C (2013) 'Flying Markets' Activating Public Spaces in Amsterdam. Built Environment 39(2):245–260

Katircioğlu ST (2010) International tourism, higher education and economic growth: The case of North Cyprus. The World Economy 33(12):1955–1972

Kiessl M (2012) Art Déco Architecture in Cyprus from the 1930s to the 1950s. Prostor 1(43):172–185

Knierbein S, Aigner J, Watson S (2012) Straßenmärkte in Wien und Budapest als Schauplätze des Wirtschaftens in Migrationskontexten. In: Dabringer M, Trupp A (eds) Wirtschaften mit Migrationshintergrund. Zur soziokulturellen Bedeutung "ethnischer" Ökonomien in urbanen Räumen. Studienverlag, Innsbruck, pp 93–106

Leventis P (2005) Twelve times in Nicosia – Nicosia, Cyprus, 1192–1570. Cyprus Research Centre, Nicosia

Liverpool City Council (n.d.) Homes for a Pound. https://liverpool.gov.uk/housing/homes-for-a-pound/. Accessed 03 Jan 2023

Markides D (1998) The issue of separate municipalities in Cyprus 1957–1963: An overview. Journal of Mediterranean Studies 8(2):177–204
Mars R, Kohlstedt K (2020) The 99% invisible city: A field guide to the hidden world of everyday design. Hodder & Stoughton Ltd, London
Michaelides D (2012) Historic Nicosia. Rimal Publications, Nicosia
Morales A (2009) Public markets as community development tools. Journal of Planning Education and Research 28(4):426–440
Mousset P (1971) À Chypre. Revue des deux mondes.Nov.:332–346
Muñoz-Viñas S (2009) Contemporary Theory of Conservation. Elsevier, Oxford
Nonument (2023) Bandabuliya. https://nonument.org/nonuments/bandabuliya/. Accessed 03 Jan 2023
Offer A (1997) Between the gift and the market: the economy of regard. The Economic history review 50(3):450–476
Patrick R, Bater J (1976) Political geography and the Cyprus conflict. University of Waterloo, Department of Geography, Ontario
Podder AK, Hakim SS, Bosu SP (2018) Ordinary heritage. ArchNet-IJAR: International Journal of Architectural Research 12(2):334
Pottie-Sherman Y (2011) Markets and Diversity: An overview. MMG working paper 11–03. Max-Planck-Institut zur Erforschung multireligiöser und multiethnischer Gesellschaften, Göttingen
Rickles CJ (2019) What is leftover: The residual space. Dissertation, Georgia Institute of Technology
Salvator L (1983) Levkosia, the capital of Cyprus. Trigraph Limited, London (Original work published 1873)
Scoppetta C (2010) Gli "in-between spaces". Elementi caratterizzanti della metropoli contemporanea. Papers Cittalia, Roma
Solà-Morales I (1995) Terrain Vague. In: Davidson C. (ed) Anyplace. MIT Press, Cambridge, pp 118–123
Sylvia Ioannou Foundation, Harokopio University (2022) Kitchener's survey of Cyprus, 1878–1883. https://kitchener.hua.gr/. Accessed 03 Jan 2023
Theodotou Y (2016) Markets and Their Social Space. Dissertation, University of Nicosia
UNDP (2013) Bandabuliya: Not just another market! https://www.undp.org/cyprus/publications/bandabuliya-not-just-another-market. Accessed 03 Jan 2023
UNHCR (1995) A Vision for the Future of Nicosia: A Partnership towards the Rehabilitation of Chrysaliniotissa and Arab Ahmet, Nicosia
Watson S (2009) The magic of the marketplace: Sociality in a neglected public space. Urban Studies 46(8):1577–1591
Watson S, Studdert D (2006) Markets as sites for social interaction: Spaces of diversity. Policy Press, Bristol
Willinger S (2007) Bilder von Aneignung und Gebrauch–die soziale Produktion urbaner Freiräume. Informationen zur Raumentwicklung 12(11):731–739

Civic Water. Bridging Culture, Nature and People

Samidha Pusalkar, Norma Camilla Baratta, Massimo Izzo, Danila Saulino, and Filippo LaFleur

Abstract The essay explores the potential reuse of traditional urban water infrastructures in Nicosia, with reference to the walled city. The political and physical partition of the city has not only pushed for the material decay of the built environment inside and bordering the Green Line, but also the abandonment of various common infrastructures, such as the water system. This is even more critical in Cyprus, an island suffering from severe water scarcity. Hence, Nicosia's water legacy has been studied, mapped, and analyzed from its patrimonial perspective, with attention to its tangible and intangible manifestations. This research enables the potential to connect the two communities by virtue of a social and renewable use of water. To promote this vision, the research puts 'water heritage' as a theme under which community-based initiatives and the reintroduction of nature as a service in the urban fabric are proposed with different social and cultural target actors, based on the methodology of "stapling strategies".

1 Introduction

This research investigates how urban water infrastructure, and the related water heritage can act as a compound theme for bridging Nicosia's Greek Cypriot and Turkish Cypriot communities. The project started off by recognizing and then

S. Pusalkar (✉) · N. C. Baratta · M. Izzo · D. Saulino · F. LaFleur
Department of Architecture and Urban Studies, Politecnico Di Milano, Milan, Italy
e-mail: samidha.pusalkar@polimi.it

N. C. Baratta
e-mail: normacamilla.baratta@polimi.it

M. Izzo
e-mail: massimo.izzo@polimi.it

D. Saulino
e-mail: danila.saulino@polimi.it

F. LaFleur
e-mail: filippo.lafleur@polimi.it

© The Author(s), under exclusive license to Springer Nature Switzerland AG 2023

A. Buoli and O. C. Ţiganea (eds.), *Territorial Fragilities in Cyprus*, Research for Development, https://doi.org/10.1007/978-3-031-36076-3_10

analyzing the public awareness and engagement about the walled city's historic water network, despite the mutual dependence of communities on water access, as well as the critical water scarcity level. This knowledge basis provided the background for the design of three operational phases for enhancing a better water management toolkit for the local authorities and a public awareness-building for Nicosia's citizens.

Water heritage stratification has followed Nicosia's changing rulers, such as the Ottomans (aqueducts, ablution in caravanserais and mosques, hammams) and the British (water retention tanks, water towers and street taps). Water is also at the heart of a geostrategic problem of dependency on the continental powers of Greece and Turkey. During the 2008 financial crisis, the Republic of Cyprus had to import drinking water from Greece using tankers, while restricting household supply by 25–30% (European Environment Agency 2009). Currently, a water pipeline from Turkey serves the de-facto Turkish Republic of Northern Cyprus (Park 2020). Hence, considering the effects of the climate crisis and socio-cultural territorial fragility at different scales in Nicosia and Cyprus, which also include the increasing water scarcity and the gap between the Northern and Southern communities, 'water' as a theme was carefully chosen for this research. This is evident with a palingenetic pattern that was seen with the extreme rainfall events due to climate change which have recently caused the collapse of one Orthodox religious symbol in the buffer zone (St. James church) (Andreou 2021; Cyprus News Agency 2021).

To explore the important aspects of water for Nicosia, a detailed survey and analysis was done, in September 2021. Along with this, research publication by Lau (2012) and comments made by Dr. Arch Nasso Chrysochou during the Nicosia visit also proved very beneficial for this study.[1] These led to confirming the disposition of the existing tear in the historic and current water network in Nicosia. With 'water' as a multidisciplinary strategic domain for both communities of Nicosia and Cyprus as a whole, it became a potential asset and a tool for tackling other related issues as well, such as environmental sustainability, economic development and bridging the divided communities. The proposed phases in this research, acting as pilot projects, indicate having community involvement as a driving factor, which makes it a campaign 'by people, for people'.

The essay follows a structure by first discussing the issues and history of water in Cyprus and then doing the same with Nicosia's walled city as the main focus. It also discusses the water heritage, its socio-cultural, environmental, and communal aspects, which are then followed by the proposed project phases (Fig. 1).

[1] Nasso Chrysochou is Associate professor of architecture at the Frederick University in Cyprus. Since 1992 she has worked as a conservation architect on innumerable projects of conservation of historic buildings and monuments in Cyprus. She has been twice awarded the Europa Nostra Award for conservation of monuments in Cyprus. Her involvement in the "Territorial Fragilities in Cyprus" workshop and Nicosia field visit are explained at the initial sections of this volume.

For more information on her research and projects, please refer to: https://www.frederick.ac.cy/en/about-us/faculty-staff/faculty?view=page&id=108&lid=586.

Fig. 1 Fountain of Es-Seyyid Mehmet Emin Efendi. Photographed 1945. *Source* Leventis Museum archives. (Lau 2012, p. 6)

2 Thematic Background Study of Cyprus

In the following section, aspects such as issues of water scarcity and a detailed analysis of the history and legal frameworks of water development are discussed, in the context of Cyprus Island.

2.1 The Issue of Water Scarcity, and the Option of Stormwater Collection

Water shortage affects Cyprus as a whole, regardless of the enduring political divide. The island suffers from Europe's most severe water scarcity problem (Eurostat 2015, in Zachariadis 2016: 2). On a larger scale, southeast Mediterranean Europe is projected to be a global hotspot of desertification and sea level rise. Zachariadis (2012, 2016) identifies three main vulnerability aspects related to the water element: (a) fall in freshwater availability, (b) overstressed natural groundwater resources of the island, with the lowest annual freshwater resources per capita in the EU, and (c) extreme weather events, such as dry spells and cloudbursts.

We conducted a spatial time-series analysis of rainfalls in the 1981–2020 period to integrate Zachariadis' study and situate our water heritage vision for Nicosia in this framework of ecological crisis. The geophysical data source we collected and processed is the harmonized CHIRPS 2.0 rainfall database (Climate Hazards InfraRed Precipitation with Station data) (2021a). It is a stack of raster layers reporting the average monthly rainfall millimeters over the last 40 years. The estimation combines global satellite sources with local meteorological sensors at a 5 km resolution.

A complementary data source is the EU-Hydro database (© European Union, Copernicus Land Monitoring Service) (2021b). It is a photo-interpreted network including water bodies (lakes, wide rivers, e.g., the Pedieos), minor streams, and canals. The statistical language R (R Core Team 2020) was necessary to automate the sequence of spatial reprojections on a uniform system of coordinates, given the size and number of layers in CHIRPS 2.0. As a result, we were able to pivot the monthly data to seasonal data averaged over two periods of 20 years, 1981–2000 (t_1) and 2001–2020 (t_2), keeping the wet and dry seasons separate for comparison (autumn–winter vs. spring–summer).

After aggregating the data on an average seasonal basis, we now observe the general contraction of rainfall from t_1 to t_2 in the dry seasons (Fig. 2, right). The shrinkage of rainfall extent is mostly visible around the Troodos Mountain (southwest Cyprus), the origin of the main hydrographic flows in all directions. On the contrary, the wet seasons have seen an intensification of rainfall in the same area over the 40-year period (Fig. 2, left). As shown by the greening tiles around the capital, the rising average precipitation levels have propagated to the Green Line in a "donut shape", highlighting the condition of Nicosia as a warming inland city. While the surrounding regions have registered at least a mild increase in humidity, Nicosia persists in a central rainfall depression basin (the hole of the "donut").

Overall, the decline in average precipitation is a trend parallel to the growing frequency of extreme stormwater events, leading to periodic floods in the city of Nicosia. Recalling the palingenetic cycle, we mentioned at the beginning, the inundation events are concentrated in the walled city's Buffer Zone, namely the Paleo-Pedieos. Records since the fourteenth century confirm that the hardest-hit part of the city is where the Venetians diverted the river to the north, away from its natural entrance at the Paphos Gate (Charalambous et al. 2016). So, despite the forceful breakup in the 1970s, the north and south sides are path-dependent on the same historical decisions. Adding to this path dependency, the elevation profile also descends to the north, indicating a higher flooding risk in Turkish Cypriot Nicosia. The flood-proneness of Northern Cyprus, on the other hand, does not relieve the southern part of key environmental protection duties. Considering topography once more, Greek Cypriot governments are accountable for the urbanization and erosion dynamics possibly occurring at the foothills of Troodos, including the overland and subterranean water runoff that goes all the way to the city through the hillside.

In the arid and low-precipitation climate of Nicosia, storms become crucial chances to gather freshwater supplies and put the resource back into circulation. The tropical heat waves and the growing demand for freshwater will exacerbate the competition between climate-sensitive sectors, especially agriculture and tourism, which are integral to the island's economic base (Park 2020). As observed in Park (2020), local farming activities are exhausting the available water wells due to intensive use practices. In our experience as visitors, drinking water was stockpiled and served in plastic bottles, with no mainstream reliance on the few active fountains.

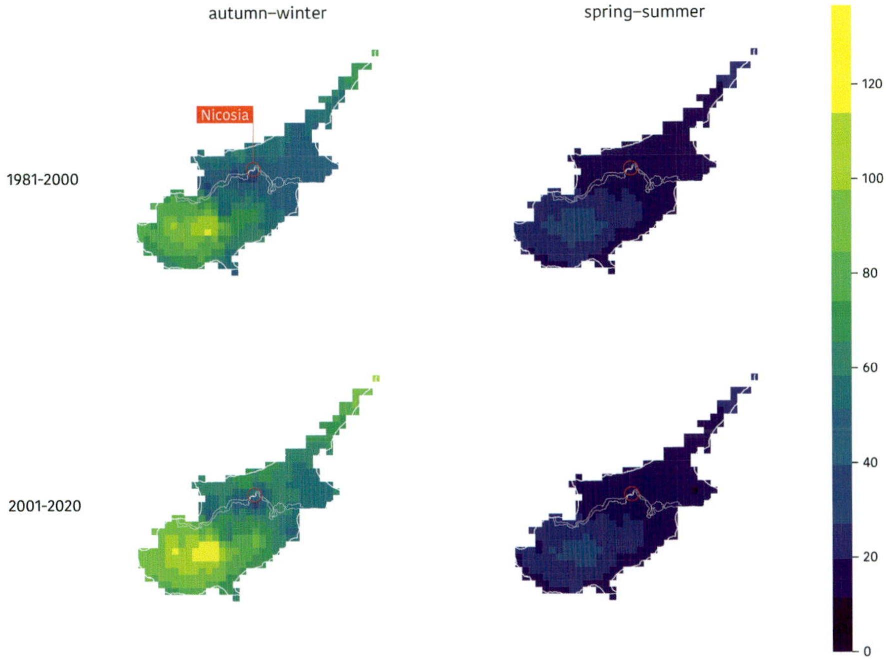

Fig. 2 Average monthly rainfalls by season (mm). *Source* Elaboration of CHIRPS 2.0 in R-spatial. Climate Hazards InfraRed Precipitation with Station data

At the same time, it is striking to see how Nicosia is also home to beautiful water-savvy architecture from the past. Aside from street taps and fountains, civil and religious examples include the British water towers and Ottoman structures like the cisterns in the Bayraktar Mosque and the active hammams. A cistern functioned with rainfall runoff storage in the absence of aqueduct pipes (Mays et al. 2013). Particularly, we noticed that the Ottoman caravanserai in North Nicosia, the Büyük Han, is reminiscent of this crafty adaptation to the water cycle. Han's rooftop was equipped with rainwater drain-off tubes, similar to gargoyles on top of gothic cathedrals. In this way, the master builders left a minimum technical space to let the rainwater channeling perform smoothly, while visitors, patrons, and merchants found shelter under the building's arched porticos, named "revaks" (Germerli 2016: 45).

Treasuring these insights, the dramatic problems of flooding and poor freshwater supply seem to tell the story of a lost connection with the water element. In truth, we abandoned this blunt view during the field visit.[2] We were surprised to learn from Agathoklis Agathokleous,[3] a civil engineer at the Department of Technical Services, that Nicosia has a *sanitary sewage system*. In the opposite *combined system*,

[2] Undertaken during Sept 2021 (Kindly refer to the initial sections of this volume which provide more details).

[3] During the field visit, it was necessary to also visit and gather in-detailed information on the existing water systems and networks. Hence, assistance was sought from the Department of Technical

rainwater and wastewater flow into the same pipe, usually headed to sewage treatment plants (Brombach et al. 2005). In terms of resource efficiency, the benefits of a sanitary system far surpass those of a combined system, in that, rainwater is collected separately from wastewater. Rainwater pipes are linked to the street and roof drains and keep the water clean to directly recharge the natural bodies around. As we verified on the spot, the Pedieos itself is fully integrated into this sanitary system.

Together, these factors suggest Nicosia's potential as a testbed for coupling decentralized stormwater reuse systems, such as Nature-Based Solutions (NBS) (Faivre et al. 2017), with efforts to preserve water heritage. NBS are not meant to be fancy stand-alone greenery. On the contrary, they would develop an existing sanitary system to help retain the urban groundwater reserve and encourage "small water cycles" (localized rains) (Widows 2016). Few idle open spaces would provide room to steer the local evapotranspiration process and soften the heat island effect, which in Nicosia is exacerbated by the walled city's dense and granular fabrics. So, other than the aesthetics of NBS, one could appreciate a mixed natural and artificial system for being inspired by the historic water heritage wisdom, as well as for tapping the "donut hole" identified in the rainfall analysis.

2.2 *Water Management in Cyprus*

In response to the acute recent droughts in Cyprus, the Government encouraged a series of management measures to better control and satisfy users' demands. Those measures include water rationing, an increase in public awareness of water conservation methods, and water pricing for the improvement of water use efficiency and water saving (Sofroniou and Bishop 2014). Until 1967, the main water resource for Cyprus was rainwater. Being an island, rainwater is the main resource that supplies ground and surface water bodies. However, it is a very uneven resource, with high concentrations in the mountainous areas, while the lowland and coastal areas suffer from water drought. The difficult environmental scenario in Cyprus, aggravated by recent climate conditions, has pushed toward the use of alternative water resources such as seawater desalination processes and wastewater reuse. Yet, these resources are insufficient to meet water demand, which has become increasingly demanding in terms of quantity and quality over time. In addition, the socio-economic development of the last 40 years, which has seen the arrival of numerous new residents and tourists, has further aggravated the problem of water scarcity in Cyprus. The management of such a limited resource becomes strategic for the development of Cyprus as well as for cooperation between the two communities. In order to understand the problem of water heritage that our project is dealing with, a brief outline of developmental stages, policies, and authorities with respect to water management is presented in this chapter.

Services. For more information kindly refer to contacts at: https://www.nicosia.org.cy/en-GB/municipality/services/technical/.

2.2.1 History of Water Management

In Europe, complex hydraulic systems began to be built in the eleventh century to support population growth (Magnusson 2001). At that time, Cyprus was a Venetian protectorate (beginning in 1473) and there are no island-wide population surveys available before this period (Lau 2012: 23). In addition, as stated by Lau, after the Ottomans conquered Cyprus almost all Venetian administrative records were lost. However, information about water management in Cyprus during the Venetian age can be deduced by comparing Cyprus with similar territorial contexts such as Crete. During the Venetian rule, both islands experienced growth in material well-being that attracted people from rural areas to the growing urban centers. Therefore, to support this growth, urban development would have necessitated a well-functioning water supply system (Lau 2012).

When the Ottomans took over Cyprus, they inherited a well-developed system of agriculture that was based on wells, irrigation channels, and water mills that they reinforced and expanded during their rule. Key was the role of the *evkaf* institution, which was a specific form of collective ownership administered by the state, based on the evkaf rules and shar'ia law (Lau 2012). The existing water infrastructure supplied water for religious, economic, and social services institutions as well as for private water systems. Moreover, Ottoman foundations provided funds not only for the construction of water systems but also for their repairing and cleaning, and established salaries for the related workers (Yildiz 2009, in Lau 2012: 33). As pointed out by Lau, Islamic societies developed traditional principles of water management that became incorporated into religious law and then applied in Cyprus. The Ottoman government established that access to a clean and adequate water supply was a basic right of individuals and communities.

Water management significantly improved under British control from 1878 through accords with the central government of the Ottoman Empire (also called Ottoman Porte) and remained until Cyprus became a republic in 1960. The British gained increased regulatory power after the official annex of Cyprus as a crown colony in 1915, which paved the way for administrative changes. However, the decline of the relationship between Britain and the colony hindered efforts to modernize the island's collective infrastructures. The first rebellion against British control occurred in 1931, resulting in restrictive political changes that barred Cypriots from engaging in the affairs of the state for another three decades of colonization.

From 1960 until the Turkish military operation, it was a crucial period for the study, planning, and design of water conducts, as well as for the planning of short- and medium-term strategies to meet the increasing demand for water in Cyprus. Technical and financial studies were conducted among the existing waterworks while new dams were built in some of the main cities interested in the tourism sector (i.e., Ayia Marina, Argaka, Lefkara, and others). In addition, projects to support rural areas were also developed. During 1878–1960, the water supply in the city was under the same remodeling dynamics as the rest of the capital's municipal infrastructure. Water distribution tanks were created for apartment complexes to support public fountains. Municipal officials also erected extra public fountains in community facilities such as schools and mosques. Lavatories were also provided in contemporary residences for most of the population, but tanks were under use all over the city till the current sewage system was introduced in the early 1970s.

The period between the Turkish military operation and the accession to European Union (1974–2004) was marked by significant achievements in the field of water development (WDD). Waterworks were expanded to improve the storage of larger quantities of water for various uses. To this end, the collaboration between international organizations and institutions such as the International Bank for Reconstruction and Development (IBRD) and the European Investment Bank (EIB), and with the collaboration of competent services of the United Nations (FAO) was key. During this period, projects for the development of rural areas were carried out, and the use of unconventional water sources was introduced. In addition, importance was given to raising public awareness towards water. Upon the accession of the Republic of Cyprus (RoC) to the European Union, the Water Development Department dynamically promoted the harmonization of the management of the country's water resources with the relevant European directives and regulations. However, at the same time, the water shortage conditions continued to aggravate due to the increasing requirements for drinking water from the Government Waterworks and the parallel reduction of natural water resources.

2.2.2 Legal Framework of Water Management (in the Republic of Cyprus)

Since 1896, the water management of Cyprus has been the responsibility of a Branch of the Public Works Department competent in the fields of irrigation and drinking water. In 1939, it became an autonomous Department called "Department of Water Supply and Irrigation" whose name in 1954 was changed to "Water Development Department", which is still the name to date. Upon the establishment of the Republic of Cyprus in 1959, the Department came under the Ministry of Agriculture, Natural Resources and the Environment, and it was gradually re-organized into various departments (such as the Ministry of Health, the Ministry of Agriculture, the Ministry of Interior, the District Water Boards, and the District Sewage Boards) and in five District Offices (in Nicosia, Limassol, Paphos, Larnaca and Famagusta). The aim of the subdivision was the decentralization of authorities, but it also caused the fragmentation of responsibilities in water management which is recognized as one of the main causes of the ineffectiveness of water management.

On one hand, the fragmented framework allowed the division of duties among the water development and management, on the other hand, it generated a composite legal framework where relevant actions were faced with a great level of bureaucracy and ineffectiveness (WDD).

In order to improve the management of a very limited resource and to ensure the sustainability of the water sector in Cyprus, it appeared necessary to review the legislation and concentrate the duties in a single. The idea was to work towards an integrated management of the water resources of the country, thus creating an

Integrated Water Body. Despite this idea has been never finalized due to legal obstacles, a new Integrated Water Management Law (79(1)/2010) was enforced on 15th November 2010. The approval of the Integrated Water Management Law by the Parliament of Cyprus meant a historical achievement for the national development and towards a more sustainable management of the water resources in Cyprus. In addition to consolidating all previous water management laws, the new Integrated Water Management Law (79(1)/2010) addressed the issue of fragmented responsibilities in water management transferring full responsibilities to the Water Development Department (WDD). Further, the new law established the Advisory Water Management Committee (AWMC).[4] The AWMC was composed of state services, public utility organizations, organized vocational groups, and water user organizations. It aimed to study the policy proposals put forward by the Minister of Agriculture, Rural Development, and the Environment, review those, and then submit related comments to the Minister and the Council of Ministers prior to decision-making. Today, all decisions concerning water policies are the responsibility of the Council of Ministers. As pointed out in the 2020 report by the Organisation for Economic Co-operation and Development, this reflects the strategic importance of water management in the context of extreme water scarcity (OECD 2020). Local authorities have limited technical and financial capacity whilst major cities such as Nicosia, Limassol, and Larnaca, are served by Water Boards that have more technical capacity and supporting finance policies (WDD).[5]

In conclusion, the Government of the Republic of Cyprus significantly invested in improving water networks and infrastructure to increase water supply capacity (Sofroniou and Bishop 2014) and outcomes in reducing water losses are already possible to be seen. On the one hand, this can be seen as an achievement at a broader territorial scale; on the other hand, this success achieved by the Cypriot government may obscure the very crucial need for better cooperation between the Greek and Turkish Cypriots (Lau 2012). Poor quality of the water supply is an issue that both communities are suffering from. Many public actors suggested showing this as evidence to the public in order to create new social common norms that could establish the waste of water as being socially unacceptable. Along with the effort by the government, rising people's awareness about the issue of water scarcity could be key to improving water management through sensible changes is consumer habits.

---

[4] This was initiated by European Union guidelines. For more information, please refer to: https://cor.europa.eu/en/engage/studies/Documents/Integrated%20water%20management.pdf.

[5] For more details, please see the WDD webpage: https://www.moa.gov.cy/moa/wdd/wdd.nsf/index_en/index_en?opendocument. Accessed September 2023.

3 Thematic Background Study: Water Heritage of Nicosia

In the following section, a more detailed and concentrated background study was undertaken to understand the current as well as historical narrative of Nicosia's water management. Moreover, a detailed mapping of water heritage was done during the fieldtrip[6], which is also discussed briefly with examples.

3.1 History of Riverbed and Water Management in Nicosia

The walled city of Nicosia has always been divided into two parts: in the sixteenth century, the historic riverbed of the Pedieos River was rerouted across the city and functioned as the primary commercial location for centuries (Fig. 3); there were seven bridges (one matches the current position of the Ledra Street checkpoint) that connected the two parts of the city. Once diverted outside the Venetian walls in the new moat, built for defensive reasons with respect to a possible Ottoman offensive, the old riverbed was not immediately reused.

From 1570 the Ottoman took over the city, and the old riverbed was left open, used as a dumping ground for refuse. Slowly being filled in, by 1882, during the British period, it was covered for hygienic reasons. A road emerged, and it became Ermou Street, a linear market, the heart of the city: the major commercial axis, lived by one and many communities together. Finally, about the same route became the Green Line, in 1963, after the first bi-communal conflict begun, which today separates the city and the two communities (Papadakis 2006).

"The old riverbed, through the city from Paphos Gate to Famagusta Gate, served the city's principal rain drainage system. The Greek called it Kotzirkas and the Turks Chirkefli Dere, and both words meaning the slimy torrent" (Lau 2012: 54). According to Dr. Arch. Nasso Chrysochou, the memory of the river, which has so strongly influenced the way of living in the heart of the walled city over time, is today almost forgotten.

Nonetheless, the precise overlap between the main trunk of the common sewer system and the path of the old riverbed within the city cannot fail to be somehow significant. "Rauf, don't you think there is an absolute necessity for a sewerage system? Don't you think that you need a sewerage system, and we also do?" (Demetriades 1998: 170), this is how the first of a series of meetings between Lello Demetriades, Representative of the Greek Cypriot Community of Nicosia, and Mustafa Akinci, Representative of the Turkish Cypriot Community of Nicosia, began. It is significant, in a moment of strong political conflict, that the very first bilateral meeting that took place, in the '70 s, concerned the creation of a common sewage system; the idea shared by the two representatives was that, despite the tensions between the communities, both had the right to a better quality of life (Fig. 4). On May 21,

[6] Undertaken during Sept 2021 (Kindly refer to the initial sections of this volume which provide more details).

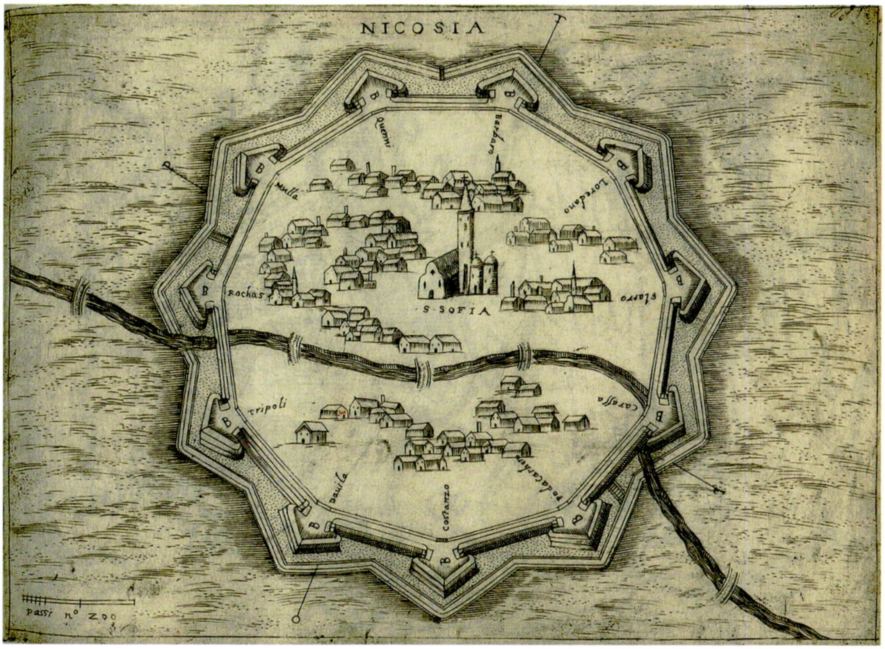

Fig. 3 Nicosia at the beginning of Venetian rule. *Source* Aikaterini Laskaridis Foundation, Camocio, G. F. Isole famose porti, fortezze, e terre maritime sottoposte alla Ser.ma Sig.ria di Venetia, ad altri Principi Christiani, et al. Sig.or Turco, novamente poste in luce, Venice, alla libraria del segno di S. Marco, [ca. 1574]. (https://eng.travelogues.gr/item.php?view=45650)

1980, the system came into operation and continues its task uninterruptedly today (Demetriades 1998).

This underground sewerage system is accepted by both sides: Nicosia, divided by ethnic conflicts above ground, is united underground through a bi-communal cooperation project. Ermou Street, which runs on the old riverbed, now finds itself running above this underground cooperation project for sewer management, also bridging part of the political divide (Papadakis 2006).

3.2 Current Water Management

Nowadays, Nicosia is served by a bi-communal Wastewater Treatment Plant (WWTP) with a capacity to treat 30,000m^3 of wastewater a day. The plant is located in the north-eastern part of the city (Mia Milia/Haspolat area). It is the largest WWTP in Cyprus and serves both the divided communities, the Turkish Cypriots, and the Greek Cypriots. Nicosia's bi-communal WWTP is part of the Nicosia masterplan sewerage

Fig. 4 'From the newspapers: The Municipalities reached agreement on the subject of the sewers of Nicosia'. *Source* From the Turkish Cypriot Newspaper 'Yeni Duzen' of the 28th January 1979 (Casaglia 2020, p. 86, Fig. 5)

and storm water collection arrangement (not to be confused with the Nicosia Masterplan—NMP), which was established by the Cyprus Government in 1960.[7] The aim of the project was to construct, operate and maintain the central sewerage system. Furthermore, the plant was seen as an opportunity to work on the city's division. It was funded by the World Bank, the Cyprus Government, and the Sewerage Board of Nicosia to facilitate the access to wastewater services for both communities but, beyond practical reasons, it was meant to be a fundamental contribution to the ongoing peace and confidence-building process.

Beginning in 1972, its execution consisted of 3 main phases. The first phase was completed in 1980 and aimed to connect main institutional buildings and key service, such as the Nicosia General Hospital. The second phase was completed in 1986 and connected the plant to the Nicosia's old town and the rest of the northern part of the city. It was funded by the European Economic Community. Finally, the third phase started in 1988 and was completed in 1995, it was funded by the United Nations High Commissioner for Refugees (UNHCR). After 1995, the plant continued to be expanded. Although, the expansion regarded only the collection area and but also the treatment plant, thus causing a strong odor that forced the administration to plan a new treatment plant.[8]

However, since 2003 the plant has shown considerable environmental problems. The population of the city sensibly increased overloading the capacity of the plant and causing unpleasant odors. At the same time, the Nicosia existing plant also needed to meet the higher European Union quality requirements. Thus, both the two communities came to discuss the construction of a new plant that could use wastewater as a resource.

[7] For more information, please refer to: https://www.water-technology.net/projects/nicosiawastewatertre/.

[8] For detailed report, please refer to: https://www.arinfuse.eu/sewerage-board-of-nicosia/.

In March 2010, the project for a New Nicosia Wastewater Treatment Plant started and was finally completed in June 2013. The plant removals contaminants from wastewater, which contributes to check the pollution levels of the river Pedieos through a physical, chemical, and organic processes, moreover, it provides water to irrigate agricultural land. In addition, the New Nicosia Wastewater Treatment Plant produces dry bio-solids used as a natural fertilizer, and green electricity from biogas. Being a bi-communal project, the funding for the plant came from different sources and in different percentage: 30% of the funding was granted for the Turkish Cypriot community by the European Commission through the EU aid program,[9] while the remaining 70% is provided by the Sewerage Board of Nicosia (UNDP 2022).

3.3 *Role of Water Heritage*

In Cyprus, water heritage has not been the object of specific preservation measures until recently. One reason for this may be the weakly established connections between water, history, and cultural identity in the collective imagination. Working on water systems in a divided city pushes to go beyond "cultural boundaries" towards a more general context, which is made of non-partisan social values.

As mentioned above, Nicosia's bi-communal sewage system operates regardless of the above-ground separation. The design and implementation of such a collaborative infrastructure began in the early 1970s (World Bank 1971), a delicate moment of clashes. This contradiction suggests the special role of water in joining communities based on the pragmatic necessity of saving a resource for all. However, well before a common sewage system, the walled city was connected by an entire network of continuous water channels. This network manifested itself in discrete pieces of traditional water architecture, visible (above ground) and invisible (underground) (Bakshi 2012).

The remnants of aqueducts, fountains, springs and vernacular rainfall retention systems are not episodic elements. Rather, they are a testimony that the city was formerly integrated into the cycle of water.

Leveraging the beauty and mastery behind such water architectures appears to be another point of convergence in a bi-communal cooperation framework, in spite of the complex separation between the two sides of the city. This suggests unveiling the historical water network as a whole and rediscovering the socio-cultural significance of it.

There is little information on the aqueducts that brought drinking water to the city. During field visits for this PhD workshop, in September 2021, it was found that two were the pipeline systems: (1) the Arab Ahmet Pasa to the west, the older one,

[9] For more information, please refer to: https://commission.europa.eu/funding-tenders/find-funding/eu-funding-programmes/support-turkish-cypriot-community/aid-programme-turkish-cypriot-community_en.

and (2) the Silahtar to the east. The latter was constructed in 1801–1803 because the former alone did not provide sufficient water to the citizens.

The Silahtar was a stone-built arched construction running from Kyrenia Gate to Famagusta Gate and supplied water to several fountains in the city. A monumental aqueduct branch with eleven arches was discovered during the demolition of a private building near Famagusta Gate (Lau 2012).

Public fountains were the direct access points to water at the neighborhood level. So, they had a prominent social relevance among water systems. Fountains represented not only works of art, but also the daily meeting places for communities, which were typically mixed at the time.

An exemplary artefact is a fountain in the middle of Samanbahca, a regularly shaped neighborhood in the North-West of the walled city.

Lau (2012) identifies three categories of fountains that are common in both Northern and Southern Nicosia: (1) *sokak çeşmeleri* (street taps), usually inset into a building wall recess, fence, or water tank; (2) *meydan çeşmeleri* (public square fountains), free-standing, monumental structures with a water tank; (3) *şardivanlar* (fountains for ablutions), located near mosques, and intended for religious rituals.

3.4 Mapping of Water Heritage in Nicosia

The water cultural heritage present within the walled city is varied and variously arranged. The main distinction can be traced back to their origins: if the ancient aqueducts, baths and most of the public fountains mainly date back to the Ottoman period, elements such as cisterns, water towers and street taps generally refer to the British period.

A greater number of elements related to the water heritage can be seen in the north of the historic city. Some of these are located nearby and interact with religious architecture, such as the fountain of the Selimlye Mosque or the fountain of Ali Ruhi in the Laleli Mosque, while others are in more public contexts, such as the fountain of Es-Seyyid Mehmet Emin Efendi and the fountain in the Samanbahca neighborhood.

The survey (Figs. 5 and 6) unites a preliminary mapping of water heritage and a checklist compiled during the field visit. Every artifact (e.g., Fig. 7a, b) has been checked against specific features to better characterize and appreciate the walled city's patrimony. The survey work format resembles ID cards and includes a very brief description of the historic artifact, in terms of period, typology, and map location. It also provides the latest status of its state of conservation with categories explaining whether its material conservation is done or nor, or if they are neglected or vulnerable or both. The same is done for its state of use, using categories such as currently in use, not in use or has changed in use. These survey points are answered based on the visual analysis undertaken (Fig. 6).

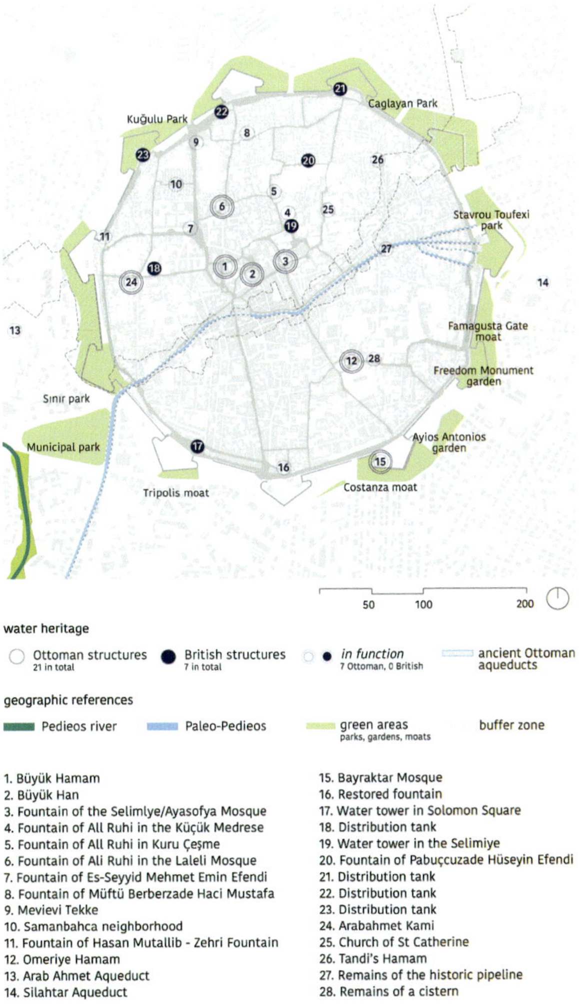

Fig. 5 Map of the existing water heritage in Nicosia. *Source* elaboration by the authors with the support of Neofytos Christou

3.5 *Brief Socio-Cultural Overview*

Water has always played a fundamental social role in every civilization, and the same is the case with Nicosia. Looking at the map showing the location and historical period of the water heritage, as mentioned above, we notice a significant majority, especially related to the Ottoman period, in the northern half of the walled city (Fig. 5). Structures such as the baths and fountains built near the mosques date back to this period.

In Islamic cultures, the purification with water is not understood only in the physical sense, to wash the body, but also as a spiritual purification which precedes religious rituals (Hudovic Kljuno 2021); the significance of the hammam (the bath)

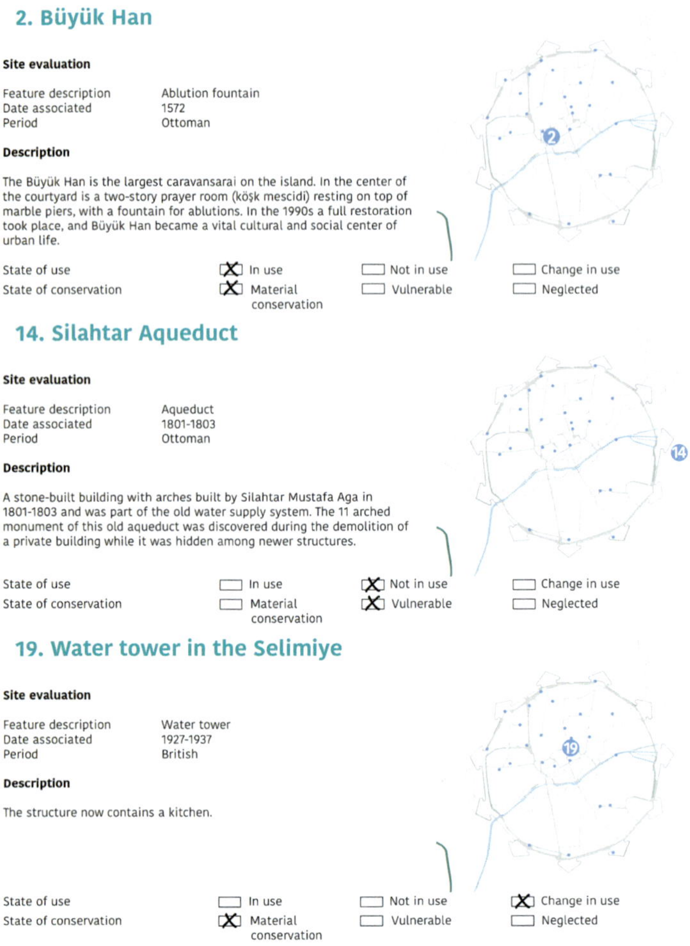

Fig. 6 Template for survey work. Cards used during site investigations as a first approach to water heritage. *Source* elaboration by the authors with the support of Neofytos Christou

Fig. 7 The Silahtar Aqueduct (14) and Büyük Han (2). *Source* photos by the authors

was indeed both religious and civic: it provided for the needs of ritual ablutions but also provided for general hygiene and served other social functions such as offering a meeting place. In the southern half of the walled city still works the Omerye Hamam, a Turkish bath that now offers a modernized spa system; instead in the north, the Buyuk Hamam[10] is the only original Turkish bath active, while the Tandi's Hamam is now no longer in use.

Also, fountains built near mosques had, and still have, where in use, an active role in religious practice: *wudu* (or ritual washing), which is performed before the prayer can be also described as partial ablution. During *wudu* certain parts of the body are washed; but it is not only physical cleansing, moreover symbolic cleansing of the mind and soul (Hudovic Kljuno 2021). We find examples of these still working fountains in the north, such as the fountain of the Selimlye/Ayasofya Mosque and the fountain of Ali Ruhi in the Laleli Mosque while in the south, the fountain of the Bayraktar Mosque.

It is interesting to understand the contexts in which the architecture of the water heritage stands today within the general socio-cultural fabric of the walled city. The division of the city, and thus the cutting up of neighborhoods and the interruption of the functional continuity of streets, led over time to the "mirroring" of activities on both sides of the Green Line, in order to integrate and re-establish functional continuity in the two municipal fragments (Pieri 2017).

Thus, the urban spaces on both sides that host the most educational and cultural activities (some schools and many museums) and those with the greatest religious and social appeal (places where the water heritage still plays an active and fundamental role) were identified. On the basis of this information, it was possible to outline a map of the walled city that highlights the relationships between these areas with respect to the water heritage, with the aim of identifying interesting contexts that may suggest a greater awareness in the definition of the project as a thematic itinerary within the city context and in dialogue with it (Figs. 8 and 9).

3.6 Community Involvement as a Driving Factor

Today, historic urban landscape not only include detached fragments of historic structures, but also larger-scale ensemble, communities, and landforms. There seems to be a widespread notion that socio-economic development activities such as awareness campaigns have a significant influence on both heritage and communities (Buckley et al. 2015). Thus, urban heritage and sustainability management are encouraged to be carried out through inclusive and innovative community involvement approaches in order to better meet the issues faced by them (Lewis 2015; Yung

[10] On the site of Büyük Hamam, the Lusignan church of St. George of the Latins was constructed from 1306 to 1309; the building was rebuilt as a Turkish bath between 1571 and 1590 during the first years of the Ottoman rule in the island. As the ground in the surrounding areas has risen over time, the hammam's door is now located around 2 m below the ground level, and the bath rooms are 3 m below.

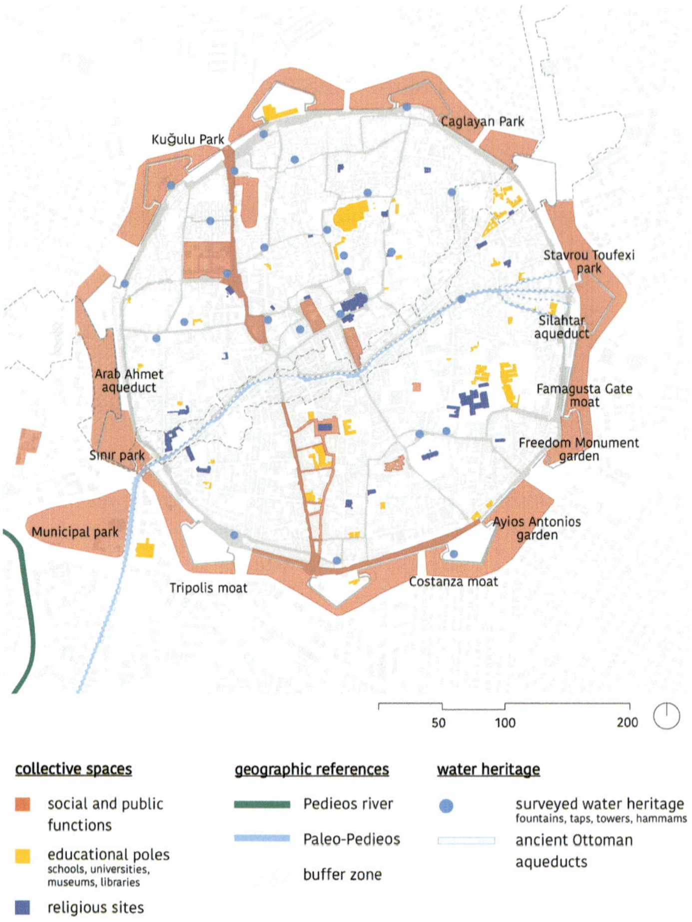

Fig. 8 Urban areas of the walled city referring to educational and social contexts. Concentrations of: schools and museums, religious sites, social and public functions. *Source* elaboration by the authors with the support of Neofytos Christou

et al. 2017). The UNESCO 2011 Recommendation on the Historic Urban Landscape (the HUL approach) emphasizes the importance of involving local communities in these dialogues within broader urban contexts by developing more holistic approaches to identify and manage the change of urban development and heritage. As noted by Poulios (2014), this community involvement can be categorized into either a *core* or a *broader* community, based on their association with historic urban landscape.

In Nicosia's case, for this project, both core communities (locals) and broader communities (public administrators, heritage experts, businessmen, tourists and funding agencies) should be involved.

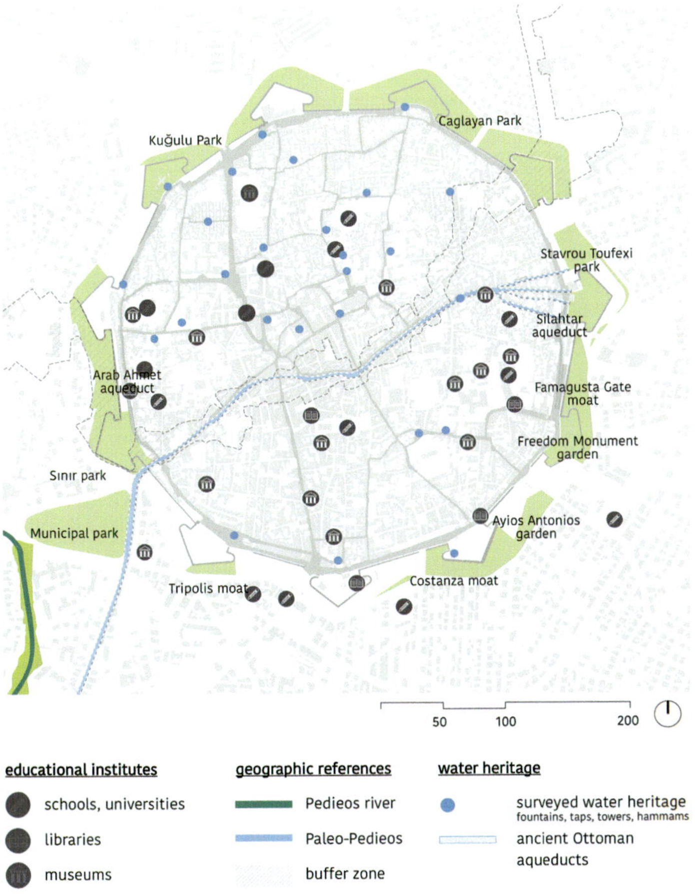

Fig. 9 Educational buildings in the walled city. While the highest concentration of water heritage appears to be in the north, most educational buildings are found in the south. *Source* elaboration by the authors with the support of Neofytos Christou

Hence, it is extremely crucial to have a clear identification of stakeholders that could be beneficial for promoting the Civic Water campaigns that we introduce and propose in the following sections. The Water Board of Nicosia, the Sewage Board, and the Technical Committee of Cultural Heritage, in agreement with the United Nations Development Programme (UNDP), have recently completed projects on conservation of water heritage fountains. Specific actors like these could step in, based on the compatibility between their interests and the initiatives denoting each phase. A number of governmental and non-governmental organizations are identified as potential actors, whose interests and resources would support our objectives in ways such as giving imaginative support and devising incentive policies to open water heritage to people and the natural water cycle. These actors are clustered into: (1) stakeholders involved in bi-communal activities, (2) technical and administrative

Table 1 Clusters of potential civic water actors

Stakeholders involved in bi-communal activities	
Home for Cooperation (H4C)	Located in the buffer zone and founded by the Association for Historical Dialogue and Research (AHDR), H4C is upfront in maintaining open bi-communal dialogue and participation
Cyprus Youth Council (C*C)	C*C has a scope similar to H4C and focuses on the Cypriot youth, also in connection with European and global youth
Pantheon Cultural Association (PCA)	The PCA brings together artists, performers, writers, musicians, filmmakers, and experimental creators
Cyprus Community Media Center (CCMC)	The CCMC works to empower civil society organizations and community groups with tools to reach a wider audience
Technical and administrative players	
Water Board, Sewage Board (Municipality of Nicosia, RoC)	These bodies provide cognitive and documentary resources, as well as monitoring support and regulation coherence
Technical Committee of Cultural Heritage (TCCH)	The TCCH is increasingly interested in the promotion and conservation of water heritage, as shown by the material fountain restoration implemented with the UNDP in 2020
Actors concerned with environmental protection	
Friends of Nature—Cyprus (FoNC), Terra Cypria	Their mission is to foster awareness and research about the island's natural and cultural richness, thereby enhancing programs for environmental protection
Educational institutes	
Schools, universities, museums, libraries	These could be variably involved in the organization of creative knowledge-spreading initiatives about water heritage history and potable / non-potable water use. The institutes could also embed science into art-led public space activations within the reach of water heritage sites. Their involvement encourages and spreads awareness among youth from a very young age, allowing them to become a responsible future of Nicosia

players, (3) actors concerned with environmental protection, and (4) educational institutes (Table 1).

4 Civic Water: Re-Connecting Nicosia's Water Heritage and Cycle

Our proposal entitled "Civic Water" looks beyond the visible division of the walled city, sparked by the sewage system as indicative of a first invisible underground cooperation. The project considers the landscape, water heritage, and the people as one entity. Indeed, it attempts to highlight that people share memory, heritage, and responsibility for the conservation of water as a vital resource. This would provide

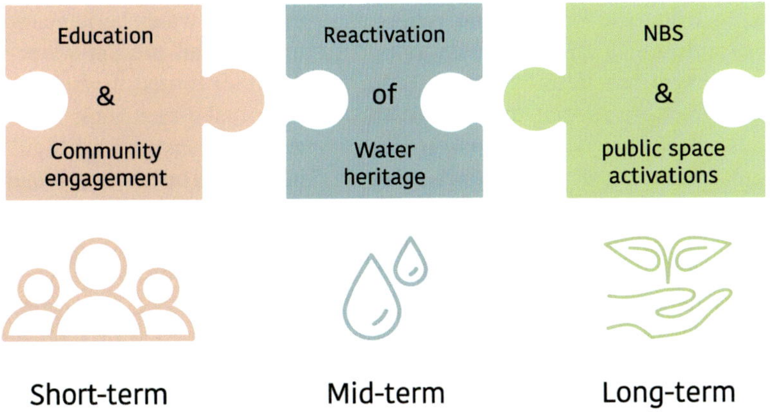

Fig. 10 Flowchart of phase works. *Source* elaboration by the authors

the walled city with a new perspective on civic water and its use—running, conscious, and plastic-free.

As noted by Saurì (2013), a way to intervene effectively in the rebalance of water demand and supply is to organize a campaign incrementally. For this reason, the proposed project is conceived to be realized in three phases (Fig. 10), where the previous phase is preparatory to the next.

The first phase is planned to be realized in a short-term period. It deals with the education and engagement of people in regard to the water heritage in place. This short-term phase serves to foster awareness and consensus over the reactivation of part of this heritage and its everyday social uses, conceived for a mid-term second phase, together with the development of a better-informed water use. Finally, the third phase develops a long-term perspective centered on a gradual renovation of open space porosities. This renovation combines art-driven public space activations with the triggering of small water cycles, and thereby making water heritage self-sustainable.

4.1 *Phase One (Short-Term)—the 'Land Vessel' Pilot Project: Treasure Hunting and Water Heritage Discovery*

The starting phase of Civic Water is inspired by the ongoing initiative by a non-profit organization, The Global Network of Water Museums headquartered in Venice, Italy. This network aims to raise awareness about the water heritage and its significance, by coordinating educational activities and campaigns, and approaching water museums and institutions worldwide.[11] The Global Network of Water Museums (2021) is

[11] A selection of "The Water We Want 2021" initiative include: (a) "Mother-Water", Aquapic Water Museum, Timişoara, Romania; (b) 'The Water We Want', Mohammed VI Museum of Water Civilization, Marrakech, Morocco; (c) 'Hourglass', EPAL Water Meuseum, Lisbon, Portugal; (d)

intended to foster the fundamental values of any form of water heritage, whether natural or cultural. It tries to create links between past and present water values and management practices by reconnecting people to all forms of water heritage, including the natural, social, cultural, artistic, and spiritual dimensions.

The historic water system of Nicosia, educational institutions, and socio-culturally significant areas together make a vibrant mosaic. These could be lenses to locally tune awareness campaigns with two parallel and communicating goals: to raise awareness of the existing historical heritage, and to develop a mind connection between city and nature, given the conditions of water scarcity that characterize the island. The walled city's water heritage is a network that can be experienced above the ground level. However, the campaign also wishes to emphasize the existence of such network beneath the ground, which is invisible yet shared by both sides of Nicosia. This concretizes the idea of giving shape to a cultural and thematic path for the entire historic center of the city, even trying to overcome existing boundaries by applying strategies for involving the public, residents or visitors.

4.1.1 Treasure Hunting Map for Students

The campaign has the potential to start educating the youth of Nicosia towards a sustainable future. Thus, the proposed pilot project for these campaigns is a treasure hunt involving the educational institutions previously identified.

This initiative gives the community a chance to reflect on a delicate and increasingly important issue, such as water scarcity, and get in touch with an urban dimension perhaps previously unknown. Exploring the city in search of 'hidden water heritage treasures' is something that students probably have already seen but never lived. Pupils live around these unknown treasures, but never give them too much attention. This small but effective initiative would make kids experience a new face of the city, not through a military division of the communities, but through an invisible network connecting them.

This would prove to be an extremely good history-based initiative, since the kids would be able to see how the walled city was and how it is now, what has changed, and what has remained similar through the interactive treasure-hunting map (Fig. 11).

4.1.2 Water Heritage Walks

In addition to the treasure-hunting initiative for kids, the project suggests that open-air museums and heritage walks could start from 'civic waters' as the main theme addressed to locals and temporary visitors of all ages.

'The Water We Created', Kahramaa Water Awareness Park, Doha, Qatar. For more details please see the Global Network of Water Museums, UNESCO Intergovernmental Hydrological Programme (2021)—https://www.watermuseums.net. Accessed April 2023.

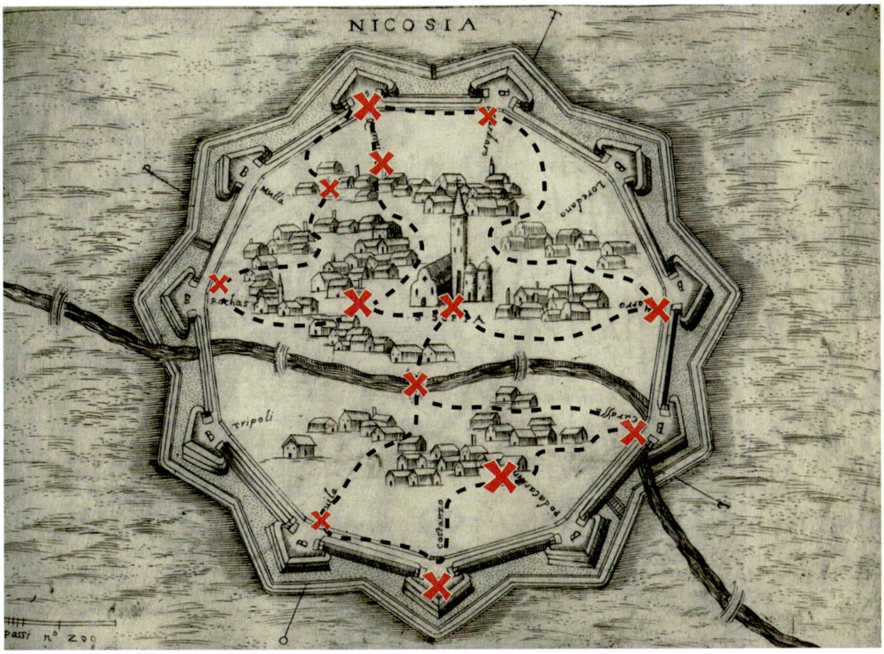

Fig. 11 How the treasure hunting map of phase 1 could look like. *Source* elaboration by the authors, based on Aikaterini Laskaridis Foundation, Camocio, G. F. Isole famose porti, fortezze, e terre maritime sottoposte alla Ser.ma Sig.ria di Venetia, ad altri Principi Christiani, et al. Sig.or Turco, novamente poste in luce, Venice, alla libraria del segno di S. Marco, [ca. 1574]

However, instead of following the traditional way of creating 'start to end' paths, people can use a map along the lines of the treasure hunt and take whichever route they like. This is done to avoid any fixed path for people, with an attempt to attract them towards these historic places even when they are busy in their day-to-day lives. This simplified idea of heritage walk could be part of the local community's daily life, and thus would help make them more aware of their heritage and civic water.

This particular project is feasible, especially with the support of the Home for Cooperation, which already works on similar initiatives. After discussing this proposal with them, during the field visit in September 2021, they not only assured a positive impact of this among the youth, but also showed interest in developing such project in the near future. It won't be an exaggeration to say that this interest is due to the lack of awareness campaigning related to Nicosia's water heritage and management.

4.2 Phase Two (Mid-Term)—'Drinking in the Rain': Water Policies, and Reactivation of Water Heritage

Community awareness-building and informed water use set the ground for enacting a better conservation of this common good and the acknowledgement of the related heritage. In the mid to long term, a growing political and social awareness could mediate the broader protection and restoration process necessary for water-related ecosystems. Reference points in this respect are targets 6.B and 6.5 of Goal 6 in Agenda 2030—Clean Water and Sanitation. Respectively, 'Support and strengthen the participation of local communities in improving water and sanitation management', and 'implement integrated water resources management at all levels, including through transboundary cooperation as appropriate' (UN General Assembly 2015).

Ecosystems can be water-related in that they are vulnerable to water scarcity, but this vulnerability is at least twofold. It means that the life of ecosystems depends on and is functional to the constant transforming of water from a state to another. A zero-waste transformation of water, its non-harmful reuse as a cooling and gathering ambiance-maker (Morello and Piga 2015), might be a bi-communal object of analysis for the actors identified previously (Table 1).

4.2.1 Water Sustainability Guidelines

In this phase, the design of community involvement could build on frequent and interactive communication channels to promote ties among interested entities, like school and university laboratories, families of the students, administrations, and local industries. This might involve the trespassing between the 'hardware' and 'software' of water, i.e., water as a public provision across city infrastructures on the one hand, and water as cultural heritage on the other. Particularly, through combining a water heritage rediscovery initiative, such as the treasure hunt, with resident education about potable and non-potable water use in Nicosia and on the island.

In developing a bi-communal actor network, the invitation of experts and local celebrities, focus groups, and even informal initiatives (e.g., games and artistic performances), are all potential accelerators of engagement and action. Importantly, they create an informed environment where citizens and institutions can reason about the variables that affect behavioral patterns in water conservation. What could follow is the exploration of opportunity windows to improve both individual and collective behaviors. Hoque (2014) identifies three groups of variables: a. socio-economic and demographic factors, like income, household size and age, schooling; b. environmental factors, like seasonal variations and sensitivity to extreme weather events, which are projected to increase in frequency and magnitude on the island (heat waves, dry spells, storms, tropical nights); c. psychological factors, i.e., immaterial values and beliefs reflected in the imagery of water (especially evident for the hammams, caravanserai, and ablution fountains on the Turkish side, but also the origin of Venus, and Beauty with her, from the sea of Cyprus). Water-savvy and behavioral change campaigns are also an occasion for presenting water use practices closer to zero-waste standards, and fostering debates on the transfer of policies to Nicosia. For instance, case studies of cities where the groundwater capacity has been recharged through the combination of individual/collective water-saving manners and techniques. These may include limits on outdoor water use, water pricing tailored to consumption styles, and refunds for water-saving devices. Advertising measures, sensors, and the extraction of control metrics would integrate the consciousness linked to the adopted behaviors.

The Civic Water actors may discuss further programs with end-consumers of water and intermediaries—household tenants, property and business managers, municipal services. For example, technical and climate-focused actors may co-design water re-evaluation programs to make property owners responsible for water management and runoff in the city. One approach is charging stormwater treatment fees based on an area's square footage. Another approach is the incentive mechanism, which could reward those who install better stormwater infrastructures (Parikh et al. 2005).

A chain of small and incremental approaches might help the municipal boards and citizens move toward more decentralized and responsive water conservation systems.

4.2.2 Reactivation of Water Heritage

The aim of this second strategy is the reactivation of the existing water heritage, which could, in a bi-communal way, be the engine and the attractor for people. To this end, after discovering that Nicosia is equipped with a *sanitary sewage system*, six existing water heritage sites have been identified for being potentially reactivated, based on the overlay to the stormwater collector, and provided that standard water sanitation filters are installed.

What we propose with this mid-term strategy is to see water heritage as an opportunity to create relational spaces that catch heterogeneous people by virtue of a common freshwater need. Here, the assumption is that the most effective way to preserve a space is to keep the space alive. In some cases, if the space in question is a terrain of division, then a new narrative of places might be needed. Places do not have a homogeneous, separate, and stable identity, but this identity is a continuous creation process (Amin and Thrift 1994). In this constant rewriting of places, people play a crucial role. The intention to bring the two communities of Nicosia closer together is far beyond the challenges of this project. What we propose is to foster the movement of people toward water heritage artifacts as a result of a reactivated flow of water.

Additionally, the proposed reactivations are located strategically on the main communication and commerce axis of the walled city (Fig. 12), stretching from the north to the south part. For this reason, it is possible to imagine these sites as a trigger for new socialization processes, and a presence in the daily life of the inhabitants of Nicosia.

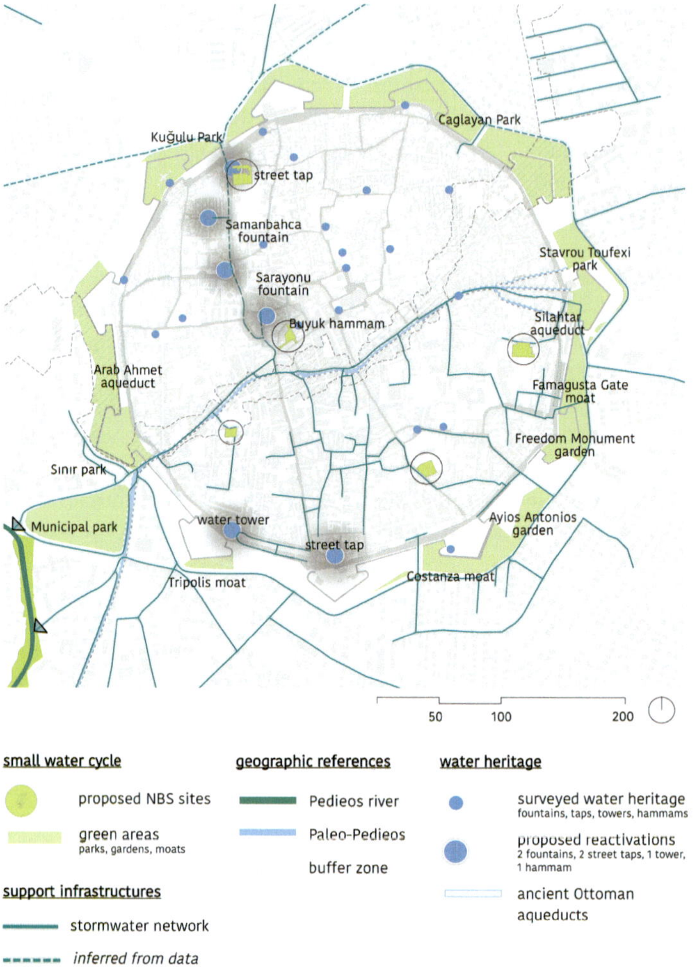

Fig. 12 Fountains proposed for reactivation in phase 2, based on the sanitary rainwater collector in place (courtesy of Agathoklis Agathokleous, Department of Technical Services, Municipality of Nicosia) and proposed locations for Nature-Based Solutions in phase 3. *Source* elaboration by the authors with the support of Neofytos Christou

4.3 Phase Three (Long-Term)—'Forecast Says Refill Included': Small Water Cycles for a Self-Sustained Water Heritage

The third phase of Civic Water focuses on strategies to activate the walled city's public spaces, and the integration of these with Nature Based Solutions (NBS), which are intended to enact, and adapt to, small water cycles. Such strategies range from stable elements, such as tree canopy and water bodies, to more ephemeral and reversible interventions, meant to reintroduce shade and permeability of soil as staples to cope with the Mediterranean climate and the rising Urban Heat Island effect. Eventually, the creation of green oases in idle spaces of the city is envisioned, as rich natural systems can flourish and sustain themselves also in hot and dry climates. Moreover, the NBS have been a strategic object of focus by various supranational institutions, which are trying to make these practices mainstream for their capacity to deliver 'services' to cities and citizens.

In the case of Nicosia, Civic Water considers NBS as functional to soften the local effects of climate change, since the island is a hotspot of global environmental crisis. Due to the straightforward connection with the topic of water, NBS are implemented as a long-term strategy for the city.

This strategy follows, structurally and functionally, the reactivation of water heritage structures in phase 2 (2 fountains, 2 street taps, 1 water tower, 1 hammam). Indeed, it is through this incremental strategy that NBS composes the last step for a robust reconfiguration of public spaces toward more livable and resilient places (Fig. 12).

According to the OECD policy analyses (2015), NBS are increasingly being recognized as the most cost-effective options to address climate change and mitigate risks in urban areas—especially water-related risks, such as flooding and the consequent loss of usable freshwater for aquifers and living beings. Particularly, Nicosia's arid and hot climate would benefit from NBS in that they are suitable to work in harmony with grey infrastructures (OECD 2015b), like the sanitary rainwater collector discovered on the field. In fact, the proposed NBS sites align purposely with such a network. Compared to the stormwater system, NBS has lower operation, maintenance and replacement costs. Furthermore, they appreciate over time with the regeneration of nature and the associated ecosystem services, while also avoiding or postponing the investment in new grey infrastructure.

NBS could be implemented in phase 3's open spaces in the forms of new materials, ranging from permeable pavements to green–blue infrastructures. These might be composed of vegetation like native shrubs and trees, to obtain a number of ecosystem services: maximizing shade and cooling; reducing the Urban Heat Island effect; replenishing the groundwater reserve by water infiltration; ensuring air purification, moisture, and carbon sequestration. All these would contribute to the general quality of the environment. Moreover, solutions for water management include the introduction of vegetated swales, rain gardens, infiltration trenches, and underground water storage. The latter has been a widespread practice of urban water management for centuries, as seen in the examples of Matera (Museo Laboratorio della Civiltà Contadina 2007) and Istanbul (Ward et al. 2017).

Furthermore, the reintegration of nature as a service for the walled city, if paired with a mix of urban design, community engagement, art, and live performances, could make the city of Nicosia a better and less fragile place to live. An assurance of these reactivations' positive impact is evident from a series of best practices and projects mentioned below.

These examples are selected in relation to the topic of water, and for their capacities to raise awareness around this precious resource for any form of life to thrive. The identified best practices revolve around spaces that are activated temporarily or permanently. From a morphological standpoint, they could be subdivided into 'paths' and 'surfaces', being in some cases permanent interventions.

Paths are crucial elements for the activation of streets through the lens of a cultural project. Along these paths, there could be moments of pause—for instance, a series of points in which awareness is directly promoted through different forms of temporary installations, whether they are media, furniture, or art. 'Adrift', the Collective Incognito project made for the Museum of Fine Art in Montreal (2021), shows a linear painting on the street representing ice melting. The objective of such a project is to activate public space through the use of art to steer curiosity over the climate crisis we are currently facing. Other examples of 'paths' include the legacy of land artist experimentation, such as Richard Long's effort in binding together open-air performances with environmental issues (Taylor 2009). In his work, a path made by walking is the essential medium through which investigating public awareness of the relationships between society and the natural world. In these cases, it is clear how streets and walking routes become part of a 'soft strategy' for public space reactivation through art, painting, and new visual perspectives. Similar to paths, *surfaces* like squares can also be transformed temporarily to communicate specific messages about the interdependencies between water and societal development.

A successful example is the network of temporary installations made for the Manifesta art biennale in Palermo, Sicily (2018). That manifestation showed the importance of art and the creative sector in activating spaces towards a greater awareness of the local territory. Specifically, the project 'What is above is what is below' is a series of artifacts designed and curated by Cooking Section Architecture Studio, which symbolically replicate pieces of ancient knowledge and design techniques to store water in cities. One of these frugal methods allows to retain moisture within arid landscapes: orange trees are wrapped in a curtain, thereby maximizing shaded grounds, and minimizing water evapotranspiration and drystone retention walls at once. Moreover, the biennale's accurate communication strategy conveyed and disseminated the project through easily understandable panels for a wide audience.

With a similar approach, 40 years before, in 1982, German artist Joseph Beuys worked with the medium of planting and de-paving public spaces as a symbolic protest act for the protection of natural resources (Strauss 2006).

Permanent interventions can also be relevant in raising awareness on the topic of water, and to make this element visible to the public. A relevant example is the renovation of a public space in a context similar to Nicosia. The project refers to the regeneration of a public square in the city of Larissa, Greece (Kolasa-Sikiaridi 2017). Here, the old water 'palimpsest' of the city is revealed through a careful urban and landscape design of the city's central square. The process of transformation involved local communities and resulted in the design of a 'sculpted river', symbolizing the lost link between the city and its river. The proposal won the important European Landscape Award as a best practice in urban transformation.

With these projects, it is evident that whether permanent or temporary, through art, performance, or urban landscape design, there is immense potential of tangible and intangible interventions in activating public spaces and contributing to the awareness of the past, present, and future relationships between water and the city.

5 Concluding Remarks

With this research, an attempt has been made to mainly emphasize two issues: water scarcity and the presence of a diffused water heritage in Nicosia. It is known that water is a multi-level strategic domain, which entwines environmental sustainability, community cohesion, and economic development as mentioned above. In the proposed campaigns, water and water infrastructure in Nicosia have been envisioned as the spatial and material domain through which a "soft" strategy of cultural reclamation and environmental regeneration is unfolded.

In the short term, awareness on the two sides is promoted, for the climate problem and the heritage value of existing water infrastructure, with the involvement and collaboration among communities. This is a springboard to the medium term, which proposes the reactivation of some of these water assets in the walled city and, in the long term, follows the identification and adaptation of some urban porosity to the restoration of small water cycles.

The water network of the city of Nicosia has not only an enormous functional value, but also a social and historical one. Through the project, an attempt has been made to recognize the relevance of Nicosia's water heritage, making something that usually remains invisible, visible with the involvement of the communities. In a complex case such as Nicosia, where the walled city is already divided both on tangible and intangible levels, using inclusive community participation as a driving factor can prove to be a key aspect of the project. Even for infrastructural reactivation, the potential is seen with the example of the field that Nicosia has for sanitary rainwater harvesting system; there have been recovery efforts on the same trajectory, albeit aimed at artifacts and not functioning in the water cycle.

This opens a window of opportunity for Nicosia and its Civic Water to potentially be picked up by local, national, and international organizations for various development initiatives. With proper funding and institutional support, the project initiatives could bring together communities to revive the memories of a shared past, create awareness of equally shared issues, and collectively face their critical climate conditions.

References

Adrift, an Installation by Collective Incognito (2021). https://www.2021.lecoeurdelile.com/en/activities/expositions/adrift-an-installation-by-collective-incognito
Amin A, Thrift N (eds) (1994) Living in the global. In: Globalization, institutions, and regional development in Europe. Oxford University Press, pp 1–22
Andreou E (2021) CyprusMail Rain destroys 16th century church in Nicosia's buffer zone. https://cyprus-mail.com/2019/02/09/rain-destroys-16th-church-in-nicosias-buffer-zone/. Accessed 6 Dec 2022
Bakshi A (2012) The Legacy of Ottoman Building in Nicosia: Hans as Spaces of Coexistence in Pre-conflict Cyprus. Int J Islam Archit 1:107–128
Brombach H, Weiss G, Fuchs S (2005) A new database on urban runoff pollution: Comparison of separate and combined sewer systems. Water Sci Technol 51:119–128
Buckley K, Cooke S, Fayad S (2015) Using the Historic Urban Landscape to re-imagine Ballarat. In: Labadi S, Logan W (eds) Urban Heritage, Development and Sustainability. Routledge, pp 93–113
Casaglia A (2020) Nicosia beyond partition. Complex Geographies of the Divided City. Unicopli, Milano
Charalambous K, Bruggeman A, Bakirtzis N, Lange MA (2016) Historical flooding of the Pedieos River in Nicosia, Cyprus. Water Hist 8:191–207
Copernicus Land Monitoring Service (2021b). https://land.copernicus.eu/
Cyprus News Agency (2021) Cyprus leaders to decide fate of collapsed church in buffer zone, Technical Committee Head tells CNA. https://in-cyprus.philenews.com/news/local/cyprus-leaders-to-decide-fate-of-collapsed-church-in-buffer-zone-technical-committee-head-tells-cna/. Accessed 15 Dec 2022
Demetriades L (1998) The Nicosia Master Plan. J Mediterr Stud 8:169–176
European Environment Agency. (2009) Water resources across Europe : confronting water scarcity and drought. European Environment Agency
Faivre N, Fritz M, Freitas T, et al (2017) Nature-Based Solutions in the EU: Innovating with nature to address social, economic and environmental challenges. Environ Res 159:509–518. https://doi.org/10.1016/J.ENVRES.2017.08.032
Germerli AB (2016) The Ottoman Han: Recovery of a Lost Typology. Syracuse University, Syracuse, NY, USA
Hoque SF (2014) Water Conservation in Urban Households - Roles of Prices, Policies and Technologies. IWA Publishing
Hudovic Kljuno A (2021) Importance and use of water in Islamic religious architecture. In: Beykent 2nd International Symposium Of Design For Living With Water. Istanbul, pp 10–15
Index of /products/CHIRPS-2.0/global_monthly/tifs (2021a)
Kolasa-Sikiaridi K (2017) "The Sculpted River of Larissa" Competing for the Council of Europe's Landscape Award. Greek Reporter
Lau M (2012) Adaptive reuse and revitalization of water heritage in Nicosia, Cyprus. Massachusetts Institute of Technology, Dept. of Urban Studies and Planning
Lewis J (2015) Preserving and maintaining the concept of Letchworth Garden City. Plan Perspect 30:153–163. https://doi.org/10.1080/02665433.2014.971127
Magnusson RJ (2001) Water technology in the Middle Ages: Cities, monasteries, and waterworks after the Roman Empire. Baltimore, Md. [u.a.]
Manifesta 12 Palermo. What is above is what is below (2018). https://m12.manifesta.org/artists-cooking-sections/index.html
Mays LW, Antoniou GP, Angelakis AN (2013) History of Water Cisterns: Legacies and Lessons. Water 5:1916–1940
Morello E, Piga BEA (2015) Experiential simulation in architecture and urban space. Ambiances. https://doi.org/10.4000/AMBIANCES.671

Museo Laboratorio della Civiltà Contadina (2007) Water-harvesting systems of Matera, from Neolithic to the first half of XX century
OECD (2020) Financing Water Supply, Sanitation and Flood Protection
OECD (2015a) Policy Perspectives: Water and Innovation for Green Growth.
OECD (2015b) Water Resources Allocation
Papadakis Y (2006) Nicosia after 1960: A River, A Bridge and a Dead Zone. Global Media Journal: Mediterranean Edition 1:1-16
Parikh P, Taylor MA, Hoagland T, Thurston H (2005) At the Intersection of Hydrology, Economics, and law: Application of Market Mechanisms and Incentives to Reduce Stormwater Runoff. Environ Sci Policy 8:133–144
Park EJ (2020) Strategy of Water Distribution for Sustainable Community: Who Owns Water in Divided Cyprus? Sustainability 12:1–21
Pieri C (2017) Selective heritage management in divided cities: focusing on Nicosia's walled city centre. Nottingham Trent University
Poulios I (2014) Discussing Strategy in Heritage Conservation: Living Heritage Approach as an Example of Strategic Innovation. J Cult Herit Manag Sustain Dev 4:16–34
R Core Team (2020) R: The R Project for Statistical Computing. https://www.r-project.org/. Accessed 15 Dec 2022
Saurí D (2013) Water Conservation: Theory and Evidence in Urban Areas of the Developed World. Annu Rev Environ Resour 38:227–248. https://doi.org/10.1146/annurev-environ-013113-142651
Sofroniou A, Bishop S (2014) Water Scarcity in Cyprus: A Review and Call for Integrated Policy. Water 6:2898–2928. https://doi.org/10.3390/W6102898
Strauss DL (2006) Beuys in Ireland: 7000 Oaks on the Hill of Uisneach. Altern Glob Local, Polit 31:101–104
TCCH Cyprus Technical Commitee for Cultural Heritage. http://tcchcyprus.com/. Accessed 15 Dec 2022
Taylor B (2009) Absurd lines, protest walks: notes on Richard Long. Sculpt J 18:
UN General Assembly (2015) Transforming our world: the 2030 Agenda for Sustainable Development
UNDP (2022) New Nicosia Waste Water Treatment Plant (ID: 00057938)
Ward KA, Crapper M, Altuğ K, Crow J (2017) The Byzantine Cisterns of Constantinople. Water Supply 17:1499–1506. https://doi.org/10.2166/ws.2017.053
Water Museum Global Network (2021) The Water We Want 2021 - Drawings. https://thewaterwewant.watermuseums.net/contest/exhibition-year-2021/drawings/. Accessed 15 Dec 2022
Widows R (2016) Rehydrating the Earth: A New Paradigm for Water Management. In: Medium. https://rwidows.medium.com/rehydrating-the-earth-a-new-paradigm-for-water-management-3567866671a2
World Bank (1971) Appraisal of the Nicosia Sewerage Project Cyprus.
Yung EHK, Zhang Q, Chan EHW (2017) Underlying social factors for evaluating heritage conservation in urban renewal districts. Habitat Int 66:135–148. https://doi.org/10.1016/J.HABITATINT.2017.06.004
Zachariadis T (2012) Climate Change in Cyprus: Impacts and Adaptation Policies. Cyprus Econ Policy Rev 6:21–37
Zachariadis T (2016) Climate Change in Cyprus. Review of the Impacts and Outline of an Adaptation Strategy. Springer Cham

Venetian Walls of Nicosia. A Palimpsest for a Common Future

Giulia Bressan, Anna Evangelisti, Paola Martire, and Livia Shamir

Abstract The overarching objective of this study has been creating symbolic and physical communication bridges between the two divided sides of the city of Nicosia. This goal was rooted in the potential of attributing new meanings to Nicosia's cultural heritage, particularly the tangible heritage of the Venetian Walls. Despite their historical significance in defining the city's history and symbolically representing it, these fortifications form a complex defensive system composed of walls, bastions, and moat structures that have always been perceived as shared heritage by the various communities that have inhabited the city since the sixteenth century. Even today, despite the persistent division, the Venetian Walls represent a rare example of shared cultural heritage among conflicting communities. To explore how the Venetian Walls' heritage component can transform from being a valuable static palimpsest into an essential tool and occasion for the construction of new spaces and shared values, this research has developed a multidisciplinary research path and a new urban vision. The aim has been to define how the walls system and its surrounding areas can be transformed into new urban green and public spaces, serving as an urban catalyst that defines the city as a whole. The original contribution of this study lies in the possibility of considering the Venetian Walls system as the object and space where the Green Line can be reconsidered and ultimately crossed, allowing for tangible connections that blend the different urban communities and cultures.

G. Bressan · A. Evangelisti · P. Martire (✉) · L. Shamir
Department of Architecture and Urban Studies, Politecnico Di Milano, Milan, Italy
e-mail: paola.martire@polimi.it

G. Bressan
e-mail: giulia.bressan@polimi.it

A. Evangelisti
e-mail: anna.evangelisti@polimi.it

L. Shamir
e-mail: livia.shamir@polimi.it

© The Author(s), under exclusive license to Springer Nature Switzerland AG 2023

A. Buoli and O. C. Ţiganea (eds.), *Territorial Fragilities in Cyprus*, Research for Development, https://doi.org/10.1007/978-3-031-36076-3_11

1 Introduction

Retracing the path that led to developing reflections and designing actions for the city of Nicosia proved to be of fundamental importance in reconsidering the primarily proposed objective: to explore possible scenarios and models of spatial intervention that could cope with the urban division, facilitate by-communal projects, and encourage the city reunification.

In line with this primary purpose, the research group has begun to explore possible aspects, or rather a specific research field to be framed in the overall work of the four groups working inside the inter-doctoral workshop "Territorial Fragilities in Cyprus", each committed to addressing a particular issue. Starting from the idea of building communication bridges between the two divided communities of Nicosia, thanks to a multi-scale and interdisciplinary approach that has characterized the research work since its first steps, our research group identified the complex system of the Venetian Walls and its role inside the city as a privileged field of action. Several reasons led the research group in this direction. From a historical perspective, the wall system represents a common heritage among the two main (ethnic) communities and the other groups living in Nicosia.

The walls are, in fact, a common architectural heritage in terms of symbolic meaning and memory since they are part of Nicosia's urban and social history. In terms of a material, tangible element since the city is divided precisely by the Green Line in two parts (Alpar and Doratli 2009).

Our general research intention concerns the possibility of considering the Venetian Walls—and their surrounding spaces—as the main object and relative spaces where the separation line could be breached to create tangible connections able to bring together the urban communities.

The main research questions that guided the early stages of the research were: *Can the City of Nicosia regain its unity through its territorial palimpsest? Is the cultural heritage of a city only a memory of the past, or can it represent the construction of new shared spaces and values?*

The fieldwork in Nicosia (September 2021) added different questions such as: *How can the Venetian Walls become a tool for defining the city as a whole?*

Thus, in-depth research work on the Venetian Walls as the main study focus was conducted both from a historical-bibliographic and an urban planning point of view. Regarding the first aspect, various topics raised in this text were examined, such as the history of the Venetian Walls, the role of a wall system within a wider urban framework, and the perception of such a similar architectural asset by people. As regards the second aspect, the research group carried out an interpretative analysis of the study materials and of the territory, defining multiple spaces and dynamics to propose a by-communal exchange.

1.1 Approach and Methodology

From a methodological point of view, the research was conducted by means of a literature review, interlinking and overlapping a series of different contributions and sources (specialized literature, lectures, videos, interviews, etc.). Above all, the frequent and rich exchange of views brought by the different backgrounds and specializations of the research group members—urban design, architecture and heritage preservation—constituted an excellent opportunity for the construction of a transdisciplinary project.

The project addressed these complex issues by making cross-contributions, exchanging knowledge, and matching objectives with colleagues in Milano and Nicosia, across the work developed by other Ph.D. candidates during the workshop. Concerning the architectural preservation field, the cultural significance attributed to the walls was central as they are recognized as cultural heritage by both communities. At the same time, from an urban planning perspective, the role of the walls has been interpreted as a key element of the urban landscape capable of becoming an experimental laboratory for new and innovative urban regeneration models.

During the first months of work (due to the pandemic), it was impossible to carry out fieldwork directly in Nicosia. Therefore, the search for useful material and documentation to initiate some early reflections was based on remote desktop research. During the study trip in Nicosia, it was possible to observe the effective state of conservation of the walls, their perception and use within the urban context, and their main challenges and potentials. The practical approach to the complex walls system made it possible to accurately evaluate what emerged during the early stages of the research; and, above all, the development and definition of a series of critical on-site considerations.

2 Historical Survey of the Venetian Walls

2.1 The Venetian Walls Over Time

From a historical perspective, the close relationship between the city of Nicosia and the Venetian Walls is evident (Calame and Charlesworth 2009: 121–142). The ancient and iconic sixteenth-century defensive system constitutes a symbol that still influences the spatial configuration and perception of the ancient city.[1] Furthermore, both Turkish and Greek communities recognize themselves in the present setting.

The construction of the first fortified enclosure with 133 turrets around the city began during Peter II's reign (1369–1382), due to the war against the Genovese and

[1] For more information on the relationships between the historic city and its fortified defense systems, see in particular De Seta C, Le Go J (1989) Le città e le mura. Laterza, Bari and Fara A (1989) Il sistema e la città: architettura fortificata dell'Europa moderna dai trattati alle realizzazioni, 1464–1794. Sagep, Genova.

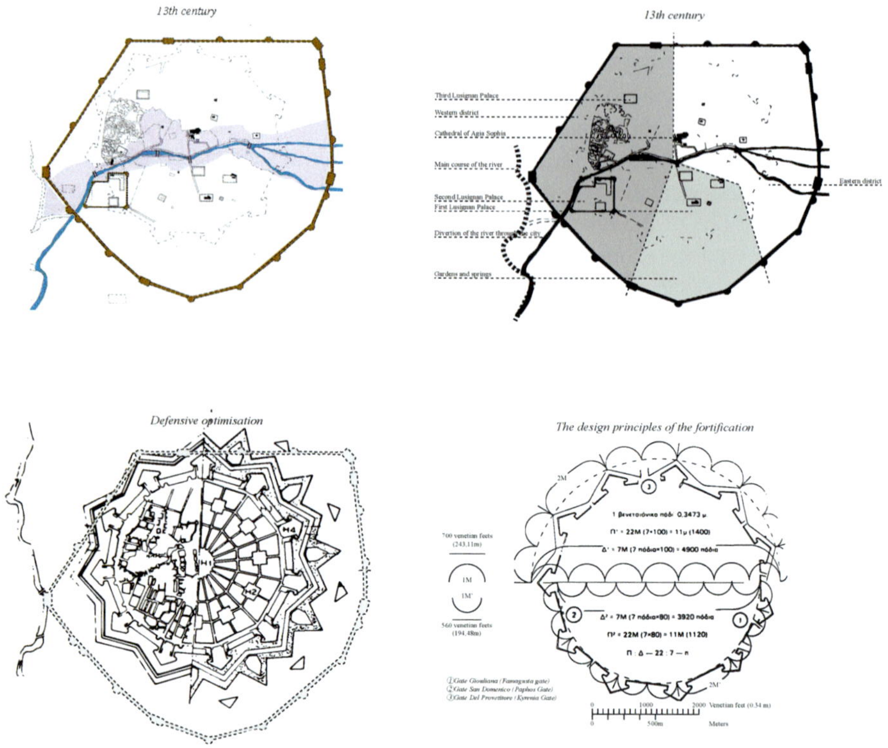

Fig. 1 **a, b** Plan of the city's fortification of the Lusignan kings related to the Venetian Walls system; Illustration of the probable configuration of the city during the Lusignan kings; **c** Savorgnan's comparison of the fortifications of Nicosia (left half) with Palmanova (right half); **d** Design and structure of the Nicosia fortification. Source: Ierides 2017: 17, 18, 22, 23, based on Keshishian (1978)

their pillage of the city in 1373.[2] "The city is characterized by the wealthy gardens and springs and palaces. The Frankish king settled in the area of the former center of the Byzantine Empire at the western part of the city that is now enriched with gothic architecture" (Ierides 2017: 17) (Fig. 1a).

Initially, the city was divided into three parts. As the illustration shows, the western part was occupied by the Frankish. On the other hand, the eastern district was inhabited by the natives. The central part was occupied by gardens and springs (Fig. 1b).

During the Republic of Venice (1489—1570), the previous fortification of the Lusignan rule was demolished to give way to an "ideal city" (Bakirtzis 2017: 174).

[2] Due to the specific theme that has been chosen for this contribution, the authors have referred only to salient events of historical relevance. Therefore, to learn more about the period of Byzantine and Lusignan domination, see Bakirtzis N., *Fortifications as urban heritage: the case of Nicosia in Cyprus and a glance at the city of Rhodes*, MAAR 62, 2017; Ierides A., *Nicosia: a review through time*, Delft 2017; Papacostas, T., *Byzantine Nicosia 650–1191*, in *Historic Nicosia*, ed. D. Michaelides, Nicosia 2012.

Fig. 2 **a** Famagusta Gate (Source: Keshishian 1978); **b** Famagusta Gate (the authors' photo, 2021); **c** Kyrenia Gate (John P. Foscolos, 1905); **d** Kyrenia Gate, available at: https://www.visitncy.com/discover/kyrenia-gate/ (Accessed: 2021); **e** Phatos Gate (the authors' photo, 2021); **f** Phatos Gate (the authors' photo, 2021)

Between 1567 and 1570, to cope with the Ottoman incursions, many buildings and fortifications of the Byzantine Empire and Frankish rule were destroyed and replaced by a more efficient and defensive walls system (Grivaud 1992: 281–306). In 1567, the Venetian's rule delegated the Italian military engineers Giulio Savorgnan and Francesco Barbaro to design the new fortifications, and the Pedieos River was diverted outside the city to fill the moat encircling the new walls.

Based on modern military architecture developments[3], Savorgnan's project was strongly influenced by Renaissance theories, which are visible in the practical examples of Palmanova (Italy), Naarden (Netherlands), and Elvas (Portugal) (Fara 1989) (Fig. 1c).

The project consisted of a circular plan[4] with three gates on the perimeter, supported by eleven heart-shaped bastions and surrounded by a moat. Specifically, the city had two diameters—the inner and the outer—divided into seven equal units. As shown in the picture, the center-to-center distance of the bastions of the Northern semicircle is equal to two miles and for the Southern semicircle is equal to two miles (Fig. 1d).

Even today, the walls retain three main entrances (Fig. 2), as follows:

1. Paphos Gate, also known as the *Porta di San Domenico*, is located on the Western side of the city wall. The road that starts from the gate leads Southwest to the city of Pathos. During the British occupation in 1878, part of the wall between the gate and the Bastion of Roccas was dismantled to create a new opening. Today, Pathos Gate sees the flags of the Republic of Cyprus and Turkey side by side, almost to symbolize a challenge.

---

[3] For a general overview of the fortification systems built by the Venetians, see in particular Boni de Nobili F, Rigo M, Zanchetta M (2016), Fortezze e baluardi veneziani. Dario De Bastiani Editore, Vittorio Veneto.

[4] About 5 km in circumference.

2. Kyrenia Gate, built in the Northern part of the fortification and known as *Porta del Provveditore*. It was restored by the Turks in 1821, who added the surmounting square building. During these renovation works, the inscription dating the door's construction to 1562, now visible above the lintel, was discovered.
3. Famagusta Gate, built in the East-southern part of the fortification between 1567 and 1570. It consists of a sizeable barrel-vaulted passage that crosses the Carafa bastion, characterized by a spherical dome raised at the center. The weapons of the Doge Pietro Loredan are located on the outer wall (1567–1570). It is also called *Porta Giuliana* or Julia in honor of the architect Giulio Savorgnan.

According to the literature review, the walls were considered the first example of sixteenth-century military architecture, on the Mediterranean coast. The wall's design brought out some innovative techniques, including the positioning of gates to the side of adjoining bastions, so they could have been more easily protected in times of siege and could have left the upper half of the wall unlined with masonry, for increasing its ability to absorb the impact from cannon shots (Bakirtzis 2017: 174–185).

During Ottoman domination in Nicosia (1570–1878), the Venetian Walls were used to provide the authorities with an instrument for controlling the city. Throughout this period, the walls retained a good state of preservation. At that time, the first restoration attempt was made; however, the attention was mainly focused on the three Gates.

During British rule (1878–1959), the colonial administration of Cyprus invested in preserving the cultural heritage of the island, due to the close relationship between the monuments and the cultural identity of the population.

Since 1903, following John Ruskin and the Society for Protection of the Ancient Buildings (SPAB)'s ideas about the medieval past,[5] the architect George Jeffery as Curator of Ancient Monuments promoted a sensible approach to preserve the walls, considered the most important monument of Nicosia (Pilides 2009). He started a series of interventions to restore the bastions and the walls system, preserving them as much as possible. When in 1927, Ronald Storrs, a new governor, came to Cyprus, Jeffery was still active in his position of Curator. He declared the whole system a monument, and in 1935 he founded the Department of Antiquities to protect and manage Cyprus archaeological and architectural heritage (Bakirtzis 2017: 178–186). The end of British rule was followed by the rise of the Republic of Cyprus (1959) and the subsequent ethnic clashes between Greek Cypriot and Turkish Cypriot communities (1962–1963).

During this period, the territorial separation between the two communities became stronger, in addition to social and political segregation. It culminated with the physical division of the city between Greek Cypriots (in the southern part) and the Turkish

[5] For a more detailed reference to theories of the art critic John Ruskin on the spiritual greatness of the Middle Ages see Ruskin J (1849) The Seven Lamps of Architecture, Smith, Elder & Co., London; Ruskin J (185–1853) The Stones of Venice, Smith, Elder & Co., London.

Cypriots (in the northern part). Another kind of "wall" or boundary was created—the Green Line (180 km)—dividing the city into two territories and dissecting the thirteenth-century Venetian Walls of the city.

Separated between North and South Nicosia, the meaning and the interpretation of rampart enclosure are changed according to the two ethnicities. As a result, the continuity of the historic circular wall has been interrupted.

A crucial moment is represented by the establishment of the Nicosia Masterplan in 1979 (Bakshi 2015). This bi-communal plan aimed to resolve critical issues and reunite the two communities.

During the 1980s, Greek and Turkish Cypriots understood the importance of the walls as a significant cultural heritage to be preserved.

The two sides of the Venetian Walls system, as will be highlighted in the section on conservation status at the present time, are mainly used for various commercial uses, parking facilities, or recreational functions.

2.2 *Meanings and Values Assigned to the Walls Over Time*

After analyzing the evolution of the Venetian Walls in relation to Nicosia's history, this paragraph is devoted to unfolding which meanings have been attributed to them over time and which cultural values the populations attributed to the walls during the different historical and political phases.

At the time of the walls' construction by the Venetians (end of the fifteenth century), and for the entire period of domination, the walls had mainly a defensive role from Ottoman attacks. Therefore, they were primarily assigned the role of protection and the value of technical and military avant-garde, engineering development, and the desire for renewal concerning the state of affairs. It is also interesting to note how, in the face of the considerable investment of resources by the Venetians for the military defense of Nicosia, the population understood the high strategic value of the city, feeling the belonging to a more defensive, military, political, commercial but also cultural system, extending from Venice to Cyprus, and affecting the Mediterranean basin (Casaglia 2020: 49–52).

Following the Ottoman conquest of the island in 1571, the coexistence between the populations and the progressive occupation of the areas within the walls began. The inhabitants still attributed a role of defense and protection and a strong urban value, delimiting the oldest part of the city to the external expansions (Casaglia 2020: 49–52). It is important to note that the Ottomans' choice not to demolish the Venetian Walls was linked to defensive matters and to the recognition of the walls as a fundamental symbol of the city. This event is of great interest if we consider that the Ottomans were gradually demolishing all the military fortifications in the newly conquered Balkan territories precisely for faster and better control of the settlements.

When English domination started in 1878, and new gates were opened, the walls gradually lost their military role, becoming more and more an urban landscape element delimiting the historic city to the external expansion areas. Relationships and exchanges between these two parts of the city (internal and external) intensified.

Starting in the 1950s, the progressive division of the city led to the interruption of the continuity of the Venetian defense system. Throughout the maximum conflict, the values attributed to the Venetian Walls between the two communities were different: "[…] for Turkish Cypriots the historic walls simultaneously signified protection and were perceived as 'prison walls', since the Turkish Cypriots minority could move outside only by passing through strict control points. For the Greek Cypriots, the Venetian Walls continued to represent the separation between the old and the new City and assumed the role of a historical monument" (Casaglia 2020: 54–55).

From 1974, the city's division into two parts and the definition of the Green Line were consolidated. The communities living separately on the North (Turkish Cypriots) and South (Greek Cypriots) sides began to attribute the same meaning to the walls: "[…] In this phase of the bordering process, the Venetian Walls definitively lost their association with the defense and became the symbol of the rich historical heritage of Nicosia and the physical and symbolic limits of the old town, a meaning shared by both communities" (Casaglia 2020: 55).

With the definitive separation of the city in 1974, both communities (Turkish Cypriots and Greek Cypriots) began to recognize in the Venetian Walls a shared past and meanings (such as defense, protection, development, etc.). Although physically separated, the two communities identify the walls as a cultural asset to be defended, preserved and transmitted, a shared cultural heritage.

Starting from this moment, the Venetian Walls are considered by both communities as a bearer of collective memory in Nicosia, where citizens from both sides recognize common meanings in them (Casaglia 2020: 55). The cultural heritage represented by the Venetian Walls is a mirror of a multiplicity of different shared values, some belonging to the most recent and dramatic history of the city, some others with the potential of becoming a tool to create a connection between the two parts of the city.

2.3 *The Venetian Walls as a Landscaping Agent*

At the end of this section relating to the analysis of the role and perception of the Venetian Walls, an interpretation of the walls as a landscape agent is proposed by adopting the analysis method on three different levels (e.g., border, cultural, and institutional landscape) offered by geographer Anna Casaglia for the Green Line, to which it is necessary to add an understanding of the Venetian Wall as a military landscape. Before exposing the primary considerations that emerged during the research, a general premise is needed. When dealing with Venetian Walls, these cannot be considered just as walls, but dimensionally and technically, as a very complex system including the wall system, the defensive escarpments, and the moat circuit. The analysis of the Venetian Walls as a unicum is illustrated below as: a border landscape, a cultural landscape, an institutional landscape, and a military landscape.

Border landscape: as already highlighted, both parts of the divided city of Nicosia today are bound by two different kinds of "borders": the Green Line and the Venetian Walls. Starting from this first level of analysis, the proposed intervention regards the progressive recovery of the Venetian Walls' unity intended as a unicum and reduction of the "weight" of the current barrier constituted by the Green Line.

Cultural landscape: regards the interpretation of the Venetian Walls as a cultural landscape, defined by Casaglia as "how the everyday use of space is marked by a divided city's state of exception, considering in particular mundane elements, museums and monuments, religious symbols, and tourism [...]" (Casaglia 2020: 148–149). Our proposal regards the progressive conservation of the walls, through reuse and enhancement projects, working with all the architectural elements of the fortifications (walls, bastions, escarpments, moat, adjacent spaces, etc.) and encouraging the daily use of public and private spaces along their route.

Institutional landscape: concerning the third level of analysis proposed by Casaglia, our proposal aims to pursue the progressive definition of a set of shared actions, including the involvement of different actors, the participation of the local community, the promotion of national and international policies, and the support to the Nicosia Masterplan's actions, especially those involving the use of the walls and the two ramparts at the intersection with the Green Line. With specific reference to this last aspect, in the formulation of the Nicosia Masterplan since its inception, the wall system has always been a focal point of great interest in which to propose new projects and concentrate multiple interests.

Military landscape: the last level of analysis on the defensive system highlighted the progressive change of meanings attributed to the Venetian Walls, which have become a landscape of memory[6]. Concerning this analysis, our proposal for the Green Line is the transition *from a divisive landscape to a landscape of shared memory*.

The examination of the defensive system as a landscape agent through four different interpretations allowed the research group to identify a series of strategic objectives for the reintegration of the walls as a unitary landscape system. These objectives—border redefinition (border landscape), creation of spaces for collective use (cultural landscape), creation of "bridges" (institutional landscape), and attribution of new meanings (military landscape)—are configured as key dimensions of the project proposals presented in the following paragraphs (Figs. 3 and 4).

[6] For a general overview of the military architectural heritage that becomes a landscape, see Damiani G, Fiorino DR (eds) (2018) Military Landscapes, Scenari per il futuro del patrimonio militare. Skira Editore, Milano.

Fig. 3 The Podocataro Bastion and moat. Photo by the authors, 2021

3 The Role and Meaning of the Venetian Walls for Nicosia and Its Urban Structure

3.1 Preservation Initiatives for the Walls

Although the conflict and the consequent city's separation process only marginally affected the Venetian Walls' system, today, the continued existence of the borderline determines a series of criticalities and risks both on a purely conservative level and on that of use.

The different preservation intervention methods between the North and the South sectors had led to the following results.

In the southern part, restoration works of the walls started in 1990 under the supervision of the Department of Antiquities (Öngül 2018: 181–188).

The sources analyzed show how, on the North side, many walls' portions were in poorer condition, even in a state of decay and crumbling parts with heavy vegetation. These conditions promoted the start in 2019 of a series of restoration works on

Fig. 4 The Roccas Bastion and moat. Photo by the authors, 2021

the collapsed portions of the Quirini Bastion.[7] In addition to these differences, it is interesting to highlight how, despite the recognition of the walls as one of the main intervention areas in the Nicosia Masterplan, the number of strategic projects currently underway—as the most famous and recent project by Zaha Hadid on the Eleftheria Square in the southern part of Nicosia[8]—is still minimal. Their punctual character does not allow them to envision the walls as a continuum and a symbolically prodromal reading of the physical process of unification of the city.

It is evident that this subdivision constitutes a limit and a possible fragility, not only in terms of conservation but also in terms of use/enhancement. Without neglecting the needs and requirements of the residents who experience daily the difficulties deriving from the presence of the Green Line, the physical impossibility of accessing the walls as a continuous path makes them unattractive even to tourists, scholars, researchers, and enthusiasts of military architecture. Finally, as a further consequence of the difficult political context, one notices the failure to include this remarkable heritage monument with unique features in international circuits of study, research, and enhancement, such as the UNESCO transnational serial website on the Venetian

7 https://www.undp.org/sites/g/files/zskgke326/files/migration/cy/UNDP_CY_FS_QuiriniBastion.pdf. Accessed March 2023.

8 https://arte.sky.it/news/zaha-hadid-architects-piazza-eleftheria-nicosia-cipro. Accessed March 2023.

defense works (Zanardo 2018).[9] The website was designed precisely to reconnect the main architectures built between the sixteenth and seventeenth centuries from the Serenissima Republic but was subsequently greatly reduced (six sites overall) also due to the uncertain political–cultural context of Cyprus.

3.2 An On-Site Overview

Several critical issues have been identified on the urban scale, affecting the city as a whole and referring specifically to the Venetian Walls system.

Firstly, a state of progressive decay and emptiness of many buildings emerges, especially those in proximity and along the Green Line. The physical deterioration of the architectural artifacts and heritage sites is added to the degradation of the surrounding environment and public areas. The city's spaces are not lived in an equivalent manner, and several areas on the Southern and Northern sides are in an advanced state of material decay and suffer a general lack of green and recreational spaces.

With reference to the walls, it was found that the city center misses their physical perception. The walls no longer present themselves as a single continuous system but are strongly disconnected due to the city division and the presence of the Green Line. This discontinuity is also recorded in elevation. The connection system between the ridge, the bastions, and the moat is interrupted or absent, making the moat itself and the spaces around the walls completely inaccessible. Even the moat is not a continuous element. On both the north and south side, some best-preserved areas are used as parking lots, football fields, or dedicated to private activities. Other areas, close to the Green Line, are left in a state of neglect and advanced degradation.

Regarding the conservation status of the walls and bastions, the walls are in an overall good state of conservation, especially in the Barbaro, Mula, Roccas, Tripoli, Costanzo, and Podocataro bastions, due to their secondary usage. It can be noted that there is a slight difference in colors between the walls of the south (lighter, yellow) and those of the north (darker, grey/black). As also Öngül (2018: 186) highlights, D'Avila, Garaffa, and Quirini bastions are among "the most attractive places of the walls, due to the buildings located on them". There is no advanced deterioration on the city's northside; recurrent and widespread decay phenomena of weeding vegetation, biological colonization, and circumscribed phenomena of erosion and lack of ashlars are present. Although there have been interventions aimed at establishing a new dialogue and relationship between the walls and the ditches—as evidenced by the intervention of Zaha Hadid on Eleftheria Square—there are still many inactive and neglected portions of the ditch, such as that of the Flatrus Bastion, due to its proximity to the Greenline. Finally, it is possible to highlight the lack of recognition of the value of the walls as a unitary system. However, the symbol of the unified walls is recurrent

[9] https://www.unesco.it/it/patrimoniomondiale/detail/480. Accessed March 2023.

Fig. 5 The walls' symbol inside the city, present on both the north and south sides of Nicosia. Photos by the authors, 2021

in both parts of the city and is visible in multiple elements, from manholes to different signs (Fig. 5).

4 Nicosia and Its Future Through the Venetian Walls

4.1 Defining the Vision for the Walled City

> The historical and architectural heritage should not be understood as a static cultural heritage, but rather as a driving force capable of bringing new creative energy to cities by promoting human development and the quality of life (Faro's Convention, art.1).

In our vision, the past becomes a tool for building Nicosia's urban future. In the context of the shared "stapling" strategies explained in the previous contribution to this volume (Pasta et al.), the defensive system of Nicosia and the project on the Venetian Walls are perfect examples of how stapling strategies can be applied to bridge the divide between different communities and promote shared identity. By expanding the common ground between the two sides of the Green Line in the walled city, our project contributes to creating a platform for engagement and collaborative activities across and beyond the border, through material upgrading, urban reactivation, and civic collaboration on both sides. This process can serve as a model for establishing connections between inter-communal planning and civil society at large, envisioning a common urban future and fostering reconciliation among different social groups. The project aims, therefore, to demonstrate how cultural and historical city centers can become laboratories to test new models of sustainable urban regeneration and lead an urban ecological transition.

The development of a new approach, combining technical, environmental, organizational and social innovation, proves that the historical heritage of the Venetian Walls will be a powerful engine for the transformation of Nicosia as a whole.

The heritage lens can focus this effort and invigorate the urban environment with creative solutions.

The project suggests bringing together the different communities in Nicosia and the two public authorities under a common goal: to improve life quality through participatory socio-cultural and entrepreneurial synergies.

The Venetian Walls system is envisioned as an emblem of the city's heritage and identity and as a platform to facilitate the activities of young entrepreneurs, creative networks, community groups and municipal services into a common green space encouraging experimentation, exchange, and dissemination of ideas through open processes and actions.

The vision for the Venetian Walls is intended, thus, to become a mechanism of social cohesion, cooperation, and co-creation in one of Nicosia's most multicultural and historical areas.

In the vision, we propose a new model of a free zone, a circumscribed area where economic and fiscal benefits will be granted. A "zona franca" is usually created to attract capital and technology and initiate a growth process. In Nicosia, the proposed free zone could have a slightly different conception: a defined space in which there is no political division that holds up and for this reason, it can become a place of mutual acceptance through daily uses and habits besides the shared value of the walls.

4.2 *Problem-Setting: De-Population and Deterioration of the Built Environment in the Old Town*

At the scale of the walled city, there is a noticeable phenomenon of depopulation and deterioration of the historical center, which is often more pronounced on one side of the Green Line than the other. This leads to a general decay of the architectural heritage and historical sites that make up the walled city. This decline is caused by the lack of social and economic activities in the city center on both sides, which creates a negative feedback loop. In addition, there is a lack of green areas and recreational spaces for the socialization of walled city users and inhabitants, resulting in a deterioration of the environmental quality.

The walled city has physical and social fragmentation issues that lead to community and facility polarization and spatial division. Focusing on the Venetian Walls, several issues serve as the starting point for the project. Firstly, as mentioned, there is a lack of physical perception of the Walls from the city centre to the outside, due to their structure and a lack of devices that enhance their presence. Additionally, there is an absence of vertical connections among the Walls' ridge, bastions, and moat, making it difficult to appreciate the Walls' full potential.

Moreover, the Venetian Walls' physical continuity is disrupted due to the city's division, and the spaces around the Walls are not easily accessible and livable. This condition leads to a general lack of use and quality of the moat's spaces and the privatization of functions along the walls. Overall, addressing these issues will be crucial in restoring the historical identity of the walled city and creating a more vibrant and sustainable urban environment.

4.3 *Cultural Heritage as a Catalyst for Urban Regeneration*

Potentialities that could build the starting point for the project's design topics were detected to effectively respond to the identified criticalities.

These can be synthesized as the need to: (a) preserve and regenerate the city's architectural heritage; (b) implement green urban networks and green areas within the walled city, to improve the number and quality of open spaces within the city center (ensuring accessibility and socialization); (c) enhance accessibility and slow mobility; (d) create opportunities for new economies; (e) improve spatial linkages between the walled city and the surroundings, and (f) enlist and recognize the Venetian Walls within an international cultural heritage framework.

Once identified these first schematic ideas of possible design topics, we pinpointed two main objectives, starting from our first research questions:

1. Regaining Nicosia's unity through its urban palimpsest.
2. Building new shared spaces and values through the resignification of the cultural heritage.

The general objectives set for the project were inspired by a lecture held by Guido Licciardi,[10] met during our workshop's intensive week. In his presentation, Licciardi elaborated on the different strategies that international organizations adopt in damaged and critical historical areas of the world to sustainably regenerate its territory, enhancing its dynamism and attractivity, and empowering communities through the creation of new jobs and the promotion of circular economy.

As he also mentions in his book (Licciardi and Amirtahmasebi 2012), there are three focal points to be considered for the sustainable development of historic cities core: livability (for locals); job creation; and local economic development. These principles were translated into a series of design and policy recommendations: the proposed project should trigger a multidisciplinary and multi-dimensional process, concerning heritage, architectural preservation and regeneration, innovative planning, with local social and economic development objectives; it should work on both sides of the Buffer Zone, inside and around the Venetian Walls, revitalizing the historic center and improving the livability; and it should enhance the perception and

[10] Guido Licciardi is a Senior Urban Development Specialist at the World Bank Group. He served the institution in operational and corporate roles, the most recent as Agile Fellow at the Office of the Managing Director for Operations. He served the Aga Khan Development Network, United Nations, Carnegie Mellon University in various countries in the Middle East.

identification of the cultural heritage of the city's walls and implement the touristic development.

4.4 Proposed Strategies and Scenarios

To make these identified criticalities and potentialities concrete and useful tools for the definition of the project, the research group determined some concrete strategies that would make explicit the intentions of the project.

The first strategy recommends rethinking the walls as a new attractive urban pole, a catalyst of intensity and diversity (intergenerational and intercultural), promoting bi-communal cooperation.

Another essential strategy regards triggering a multidisciplinary and multi-dimensional process concerning architectural preservation and innovative planning.

This process would create the conditions to rethink a possible future city reunification, working on both sides of the Buffer Zone, inside and around the Venetian Walls, while revitalizing the historical center. It would signify also a "reunification" of the green space-system of multifunctional purposes that would run along the walled city.

In addition, another important strategic intervention suggests improving the perception and identification of the cultural heritage of the city's walls and implementing their tourist development.

Lastly, an essential step for the project's development regards the institution of a possible governance model, a legal body in charge of coordinating, facilitating, and managing the Venetian Walls project over time.

Following the strategic approach and the recommendations set out above, the research group selected three main scopes for developing the project in more detail.

Considering the walls as a complex and unique system, the ridge was selected as a possible new mobility network; the bastions were identified as sites that offer different kinds of facilities or public spaces; the moat, already characterized by green areas, was opted as the site for Nature-based solutions and, more in general, landscape interventions and open spaces.

Once defined, the main strategies and scopes envisioned as contributing dimensions of a bi-communal and multidisciplinary urban regeneration process, three main scenarios have been developed, using a step-by-step and adaptive approach. Each scenario is articulated into two main phases.

The proposed first scenario called "Two Parks for Nicosia Walled City" envisions the re-appropriation of public spaces surrounding the Venetian Walls. This project is expected to activate the use of public spaces, especially the ridge and the bastions areas, and create a new urban multifunctional catalyzer along the walls' moat. The aim is to provide a new green democratic recreational space for the different communities and two new checkpoints would be established to connect the two green systems developing on the two sides of the Green Line.

The first phase (Fig. 6a, b) of the scenario will focus on the restoration of existing compatible vertical connections and the creation of a new pedestrian and cycle path network that would run along the moat on the two sides, North and South, to guarantee accessibility to the Venetian Walls' system. This phase will also promote the creation of two multifunctional urban parks, North and South, that would offer new green spaces, microclimatic comfort, cultural facilities, and economic activities. New spaces would be available for rent, especially for young entrepreneurs, creative and cultural networks, and innovation groups. This phase will rely on the provision of new urban functions that can respond to the needs of different communities, age groups, and gender. Bottom-up processes with the local communities, NGOs, and institutions will be triggered, using temporary spaces to engage people and create awareness of their tangible and intangible heritage. A new bi-communal institution, the Walled City Agency with two head offices—one for each side of the Buffer Zone—will be established to manage the project. Besides the newly formed agency, other stakeholders and users are envisioned to be involved in this phase such as associations, local committees of different social, ethnic and cultural entities within the walled city, and new potential city users.

The second phase (Fig. 7a, b) of the "Two Parks for Nicosia Walled City" scenario will focus on creating new checkpoints where the Green Line and the defensive system meet to connect the two sides of the green system along the moat. This will represent a further step towards building a unique ring along the Venetian Walls. The same stakeholders and users as in the first phase will be involved in this phase.

The second scenario, called "A Green Ring and a Linear Park" plans to unify the Green Ring into a single multifunctional urban system by gradually decommissioning the two built checkpoints. The aim is to represent a possible future integration between the green and vibrant ring of the Venetian Walls system with a linear park, a spontaneous green oasis running along the Green Line belt.

The first phase of this scenario (Fig. 8a, b) will see the overcoming of the two built checkpoints and creating a single and unified multifunctional urban Ring Park running along the whole Venetian Walls' moat, spanning from North to South. In this phase, the settlement of a single Head Office for the Walled City Foundation in the existing architectural heritage of the Flatro Bastion will be of great importance to be the first bi-communal institution within the Walled City to represent the interests of both sides of the city. Involved stakeholders and users in this phase would be: the already established Agency for Nicosia Walled City (this time with one unified Head Office transformed institutionally in a foundation), associations, local committees within the Walled City, new potential city users, and tourists.

The second phase (Fig. 9a, b) will see the integration of a new central Linear Park (former Buffer Zone), which would be the same as in Phase 1, with the addition of environmental and conservation organizations, urban planners, and landscape architects. The integration of the Linear Park with the Green Ring would require careful planning to ensure that the historic buildings and natural elements are preserved while providing new recreational opportunities for the local community and tourists.

The final step in this scenario would be the recognition of the Nicosia Walled City and its Green Ring and Linear Park as a UNESCO World Heritage Site. This would not only bring international recognition to the cultural and natural significance of the area but also promote sustainable tourism and economic development.

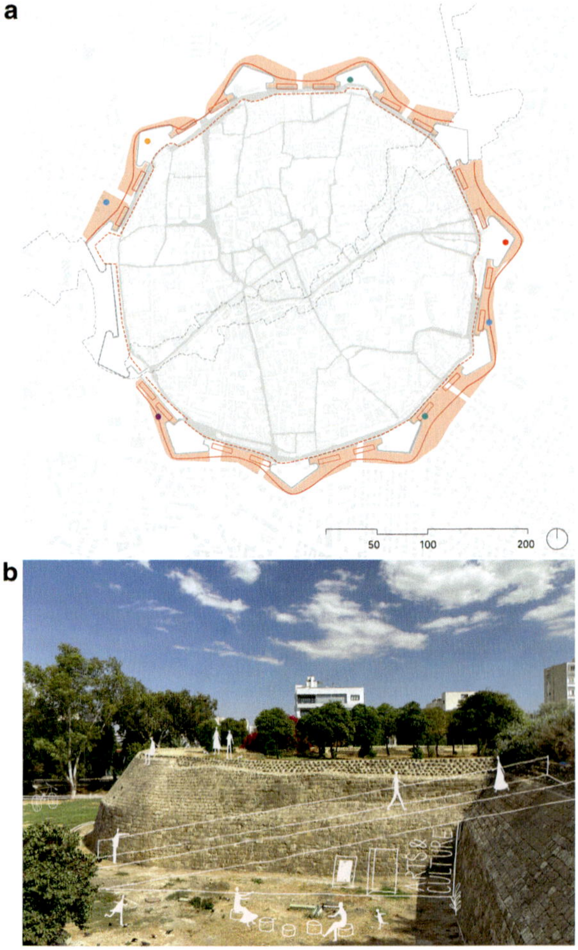

Fig. 6 Scenario A, Phase 1. Source by the authors re-edited with the support of Neofytos Christou, 2023

The "Two Parks for Nicosia Walled City" and "A Green Ring and a Linear Park" scenarios offer a vision for the future of the Nicosia Walled City that prioritizes the preservation of cultural and natural heritage while promoting social, economic, and environmental sustainability. These scenarios rely on collaboration between stakeholders and the participation of the local community to ensure that the vision is realized in a way that benefits everyone. By creating new green spaces and recreational opportunities, these scenarios offer a way to revitalize the Nicosia walled city and create a unique urban ecosystem that integrates historic and natural elements. The implementation of these scenarios would require careful planning, investment, and ongoing management, but the benefits would be significant in terms of cultural preservation, community development, and sustainable tourism.

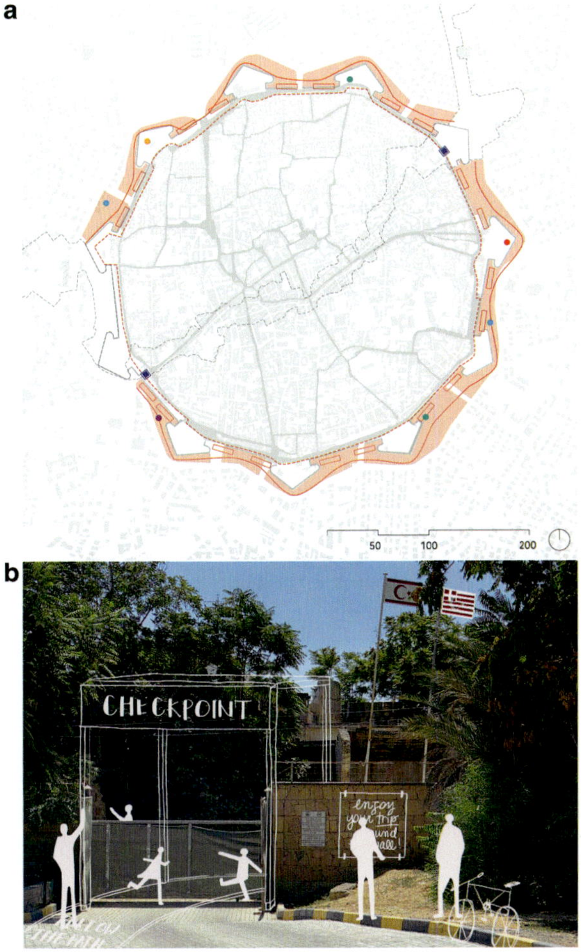

Fig. 7 Scenario A, Phase 2. Source by the authors re-edited with the support of Neofytos Christou, 2023

5 Conclusive Remarks

The research and analysis conducted on the city of Nicosia and its defensive system have revealed several significant insights into the role of cultural heritage contested urban contexts. One of the most important takeaways is that the historical built environment can play a crucial role in assuming cultural, social, political, and identity relevance in a given community. Moreover, if cultural heritage is perceived through its common values among different communities, it can also foster cross-cultural understanding and shared identities. In the case of the city of Nicosia, it became apparent that the city's heritage is the bearer of collective memories derived from the past, providing the physical platform for the construction of new meanings and shared imaginaries. Our team's decision to imagine a project starting from the historical building and the palimpsest of the Venetian Walls, therefore, became an essential

Fig. 8 Scenario B, Phase 1. Source by the authors re-edited with the support of Neofytos Christou, 2023

a

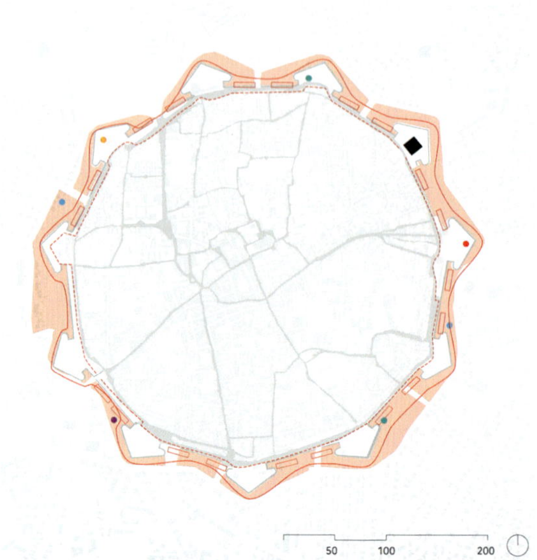

b

action to restore the city's historical identity and give new life and meaning to the walls.

It is worth noting that contemporary projects can contribute to the enhancement of historical heritage by creating new urban spaces that respond to the community's current needs. However, it is crucial to create these new spaces in full respect of the pre-existing ones. In doing so, these projects can bring new vitality to the city while preserving its heritage and cultural identity. Through the restoration and revitalization of historic buildings and structures, it is possible to promote cross-cultural understanding and shared identities while maintaining and further enhancing the cultural heritage of the community. This project has not only contributed to a deeper understanding of the Venetian Walls and their role for Nicosia, both south and north, but has also demonstrated the importance of interdisciplinary research and collaboration in the field of heritage preservation and urban planning.

Fig. 9 Scenario B, Phase 2. Source by the authors re-edited with the support of Neofytos Christou, 2023

a

b

References

Alpar AR & Doratli N (2009) Walls in Cities: A Conceptual Approach to the Walls of Nicosia. Geopolitics 14(1), p 108-134

Bakirtzis N (2017) Fortifications as urban heritage: the case of Nicosia in Cyprus and a glance at the City of Rhodes. MAAR 62, p 171-192

Bakshi A (2015) The Nicosia Master Plan: Planning Across the Divide. In: Björkdahl A, Strömbom L (eds) Governing Contested Issues in Divided Cities, Nordic Academic Press

Boni de Nobili F, Rigo M, Zanchetta M (2016), Fortezze e baluardi veneziani. Dario De Bastiani Editore, Vittorio Veneto

Calame J, Charlesworth E (2009) Divided Cities. Belfast, Beirut, Jerusalem, Mostar, and Nicosia. University of Pennsylvania Press, p 121–142

Casaglia A (2020), Nicosia Beyond Partition: Complex Geographies of the Divided City. Unicopli, Milano

Damiani G, Fiorino DR (eds) (2018) Military Landscapes, Scenari per il futuro del patrimonio militare. Skira Editore, Milano

Damgacı AY, Dağlı UU (2018) Shifting Boundaries of Divided City Nicosia through Social Actors. Space and Culture 21(4):482-494

De Seta C, Le Go J (1989) Le città e le mura. Laterza, Bari

Fara A (1989) Il sistema e la città: architettura fortificata dell'Europa moderna dai trattati alle realizzazioni, 1464-1794. Sagep, Genova

Fiore FP (ed) (2014) L'architettura militare di Venezia in terraferma e in Adriatico fra XVI e XVII secolo. Atti del Convegno internazionale di studi, Palmanova, Teatro Gustavo Modena, 8-10 novembre 2013. Olschki, Firenze

Grivaud G (1992) Nicosie remodelée (1567). Contribution à l'histoire de la ville medieval, 19 Hill G. (1940–1952). A History of Cyprus, 4 vols., Cambridge

Inançoğlu S, Özden Ö, Kara C (2020) Green Corridors in Urban Landscapes, Case Study Nicosia Pedieos River. European Journal of Sustainable Development 9(1):1-8

Ierides A (2017) Touch of Sound. Volume "Nicosia: a review through time". Dissertation, TU Delft. Available at http://resolver.tudelft.nl/uuid:268bc307-22a7-48a4-af10-83d8215b9f1a. Accessed February 2022

Keshishian KK (1978) Nicosia, capital of Cyprus then and now. Moufflon Book and Art Centre, Nicosia

Licciardi G, Amirtahmasebi R (eds) (2012) The economics of uniqueness. Investing in Historic City Cores and Cultural Heritage Assets for Sustainable Development. The World Bank

Öngül Z. (2018) Present situation of 15th Century Venetian Walls of Nicosia. In: Marotta A, Spallone R (eds) Defensive architecture of the Mediterranean. Proceedings of the International Conference on Modern Age Fortification of the Mediterranean Coast FORTMED 2018, vol 2. Politecnico di Torino, p 181–188

Pilides, D (2009) George Jeffery: His Diaries and the Ancient Monuments of Cyprus, Volume I and II, Department of Antiquities, Nicosia

Zanardo E (2018) Il patrimonio fortificato della Repubblica di Venezia come emblema di identità interculturale: per un'ipotesi di riformulazione della candidatura UNESCO. Dissertation, Department of Architecture, Politecnico di Torino

GPSR Compliance

The European Union's (EU) General Product Safety Regulation (GPSR) is a set of rules that requires consumer products to be safe and our obligations to ensure this.

If you have any concerns about our products, you can contact us on ProductSafety@springernature.com

In case Publisher is established outside the EU, the EU authorized representative is:

Springer Nature Customer Service Center GmbH
Europaplatz 3
69115 Heidelberg, Germany

Batch number: 10756376

Printed by Printforce, the Netherlands